T0146248

Bodies of Evidence

BODIES OF EVIDENCE

Medicine and the Politics of the English Inquest

1830–1926

IAN A. BURNEY

The Johns Hopkins University Press
Baltimore and London

© 2000 The Johns Hopkins University Press
All rights reserved. Published 2000
Printed in the United States of America on acid-free paper
2 4 6 8 9 7 5 3 1

The Johns Hopkins University Press
2715 North Charles Street
Baltimore, Maryland 21218-4363
www.press.jhu.edu

Library of Congress Cataloging-in-Publication Data
will be found at the end of this book.
A catalog record for this book is available
from the British Library.

ISBN 0-8018-6240-X

To Rachel

Contents

Acknowledgments

I needed a lot of help getting here, and the fact that these words can now be written testifies to the generous response over many years to what must at times have seemed like a pathological form of neediness. In the various stages and locations of thinking, researching, and writing, I have been fortunate to have had a strong community of friends, colleagues, and family to draw upon.

My journey began with visions of sherry bottles secreted in pleasantly musty collegiate offices, an image of academic existence I conjured up with the help of two undergraduate soul mates, Paul Friedland and Sarah Golin. Though the life of a practicing historian has proven more complex than we then imagined, I can at least take solace in the knowledge that our friendships have survived the retrospective recriminations about whose idea this was in the first place. My teachers and peers at Berkeley provided a stimulating environment in which to develop a more viable sense of what I was trying to accomplish. Susanna Barrows, Cathy Gallagher, Lynn Hunt, Reggie Zelnick, and especially my supervisor, Tom Laqueur, gave me examples of what humane scholarship might achieve. David Barnes, Joshua Cole, Sue Grayzel, Nicky Gullace, Page Herrlinger, Kathy Kudlick, Jeff Lena, and Sylvia Schafer offered companionship, conversation, and often a good deal more in trying times for all concerned. I am also grateful to two other institutions that provided me with a home and with intellectual and financial support while I was writing and rewriting this book. At the Institute for Advanced Study, Carla Hesse, Marianne de Laet, and Susan Kent kept my intellectual plate full, while Franz Moën and Ike Muñoz saw to it that I got more practical but just as delectable forms of nourishment. Members of the Michigan Society of Fellows and the academic community at the University of Michigan, especially Luan Briefer, Joel Howell, Martin Pernick, and Jim White, offered me an ideal environment in which to figure out what this book ought to be about. H.M. Coroner Douglas Chambers, Cathy Crawford, Saul Dubow, David Game, Colin Jones, Chris Lawrence, Jon Lunn, Hilary Marland, Iain McCalman, Francesca Nicolas, Roy Porter, Tim Screech, and James Vernon have helped to make my years of research in England productive ones. Michael Clark deserves special thanks for sharing his considerable knowledge of the history of English legal medicine with a stranger sent into the archives without a map. I also thank the Wellcome Trust for its funding of my work.

At the Johns Hopkins University Press, Jackie Wehmueller had the insight to help me out of an endless loop of revision plans by suggesting that there might already be more of a book there than I saw. Now, some years later, I am able to agree, and I thank her for waiting with such benign patience. The Press arranged for three reviews of the manuscript at different stages, two anonymous and one by Jim Epstein. Each was well judged and helpful. Linda Forlifer's careful editing made for a clearer and more effective text.

Without the unstinting and, some (me at the very least) would say, heroic efforts of four scholars and friends, I could not have written this. To recount each task they undertook on my behalf would take up far too much space; no amount of space could convey either the intellectual rigor or the generosity of spirit with which they did it. Mario Biagioli, Tom Green, Joan Scott, and Dror Wahrman—if they can bring themselves to give this one final read—will, I hope, take pleasure in what they see.

Over the life of this project, my family has not only stood—but in wonderful ways grown—beside me. My parents, Cecille and Mahmud, and my brothers, Navaid and Tarik, to be faithful to the historical record, were around even before this got started. They put up with me when they knew they knew better and shared my sorrows and my triumphs with an unwavering love. Rachel, too, has been with this since before the beginning. I turned in my French history badge for an English one so that I might be nearer her. She accepted the gesture before she or I realized its full consequences. Over the years she has selflessly given her formidable talents of criticism, creativity, and compassion. For all she has sacrificed, today she shares with me a decade of a life lived, Cailin and Rohin, a book, and my heart.

This is for her.

INTRODUCTION

This book examines the relationship between two familiar stories about the making of the modern state. The first, usually cast in terms of a rise in influence, power, and prestige, recounts the part played by scientific expertise in establishing the conceptual and practical rationale for a new, knowledge-based form of governance. Winners require losers, and this provides the material for the second story: the loser to the expert is the "public." One of the consequences of the growth of the modern state, commentators ranging from Burke to Mill to the Webbs have noted, is a decline in the traditional institutions of civic popular participation. Whether described in terms of heroic resistance, of stubborn clinging to outmoded privilege and prejudice, or of enlightened embrace of a more effective basis for social and political order, the outcome itself is not in doubt: the old will give way to the new. Of course, these two stories are linked in a series of historical and theoretical oppositions: on the one hand, knowledge authorized by local experience, institutions legitimated through historical pedigree, and efficacy measured by the value of collective self-governance; and on the other, the virtues of epistemological detachment, ahistorical reason, and efficiently disseminated knowledge at once objective and individually useful.

There are several reasons why, together, these two stories have proven such an enduring way of approaching the larger historical process of modern state formation. For one thing, like any narrative structured along binary lines, it is essentially self-sustaining. The assumed clarity of the characteristic distinctions between scientific and public modes of action tends to reproduce at a descriptive level these very assumptions and thereby to reify the conceptual terms to which they seem only to refer. Yet this approach, however satisfying, misses

a fundamental point: the world of historiographical schematics notwithstanding, these are in fact highly permeable oppositions. They
operate not so much in pristine relationships of mutual exclusivity
as in ongoing processes of interactive (and provisional) historical
definition in which neither of the ostensible poles is in itself stable
or complete.

My analysis explores this rather more complex dynamic between
expert and popular conceptions of governance from an unlikely, but
singularly suggestive, institutional site—the nineteenth- and early
twentieth-century coroner's inquest. The inquest was the anciently
constituted tribunal in English law whose primary responsibility
was to conduct inquiries into a range of specified types of death.
Since reliable knowledge of death was one of the keys to legislating
for an increasingly biological conception of "population" and since
dead bodies were themselves fundamental units of this knowledge,
the inquest's jurisdiction in one respect seemed to make it the very
type of investigative resource on which new forms of bureaucratic
management depended. It is not surprising that spokesmen for an
emerging medicolegal community waged a sustained campaign to
frame the inquest first and foremost as a tool of applied medical
inquiry. But the modern inquest was simultaneously framed within
a dynamic contemporary discourse of "historical" popular liberties.
The mere fact of its having survived from at least the twelfth century (some claimed for it an earlier, Saxon pedigree) made the inquest
appear as an exemplary embodiment of the "genius of English reform." In addition, as an open tribunal whose verdict rested with a
lay jury and whose proceedings were supervised by an elected official, the nineteenth-century inquest could be cast as a traditional
check on authority by an active and watchful citizenry.

An institution formally well positioned to take on the modern
duties of inspection and information gathering, yet at the same time
emblematic of the very participatory rationale to be displaced by the
regime of expertise, the inquest was peculiarly sensitive to the tension between the demands of expertise and those of publicity. However, both these versions—the scientific and the historical—were
innovations. The inquest was marked during this period by a struggle not so much between two self-contained rationales—one rising,

the other waning—but by an interaction between strategic visions that themselves reflected the ambiguous needs of modern governance. Plans for recasting the inquest as a bearer of usable knowledge relied on delicate, often self-subverting hybrids: innovating traditionalists, scientific democrats, popular bureaucrats. Would-be "medicalizers" of the inquest were deeply and of necessity engaged in contemporary debates about the nature, viability, and future shape of English representative institutions. It is the purpose of this book to trace out the dynamics and the consequences of this engagement.

I

Nineteenth-century inquests, to begin at the most rudimentary level, were inquiries into deaths considered worthy of inquiry.[1] This is an apposite tautology, since the precise categories of death subject to an inquest were often themselves matters of dispute according to competing standards of "worthiness."[2] Broadly speaking, however, nineteenth-century inquests were concerned with cases of accidental, suspicious, violent, or otherwise "unnatural" death. In practical terms this meant that, for the period covered in this study, 5 to 7 percent of deaths annually were made the subject of a coroner's inquest.[3]

The roughly 330 coroners for England and Wales were divided into two main jurisdictional categories, county and borough coroners. Until the end of the nineteenth century, the former were elected by county freeholders and the latter were appointed by borough corporations.[4] They served for life, subject to dismissal by the Lord Chancellor's Office only for egregious abuse of office or mental incapacity. The size of their jurisdictions and the number of inquests held varied enormously. In some districts the coronership was a full-time occupation, while in others it was a sporadically performed sideline activity.[5] From the mid–eighteenth century until 1860, coroners were paid on a fee-per-inquest basis, drawn from the local rates under the auditing supervision of the borough council or county bench. After the 1860 Coroners Act, most were paid by fixed salary adjusted every five years according to the average number of inquests held by each individual coroner. The sole formal qualification for office was to be an independent freeholder; electoral notices tended to stress local ties and general quality of character over professional identity. Throughout the nineteenth century (and, indeed, to this day), most coroners were solicitors.[6]

Inquests operated within the institutional rubric of law, but their inquiries were investigative rather than adversarial: inquests began with the fact of a death that was in some manner un- (or under-) explained, and their function was to settle on a satisfactory account of its cause.[7] As coroners were barred from initiating their own inquiries, inquests were triggered by information passed on to them from the community. The sources could range from anonymous missives from private citizens, to police, medical men, and local registrars of death.[8] Once notified, the coroner could send his officer (until the latter part of the century, typically the parish beadle; after that, increasingly a member of the local police force on permanent secondment to the coroner's office) to investigate and make a preliminary report on the circumstances, and on this information the coroner determined whether the death was a proper subject for inquest.[9]

The hearings themselves followed a common set of core procedures. Inquests began with the swearing in of a jury of twelve to twenty-four local men, summoned to serve—in theory on a rotational basis—by coroner's order. Inquest jurors were not subject to statutory qualification imposed by modern statute and could serve solely on the basis of locality and lawfulness. In practice, these panels tended to be drawn from artisans, shopkeepers, and tradesmen.[10] Once sworn, the coroner and jury retired to the place where the body lay and performed a "view of the body." Evidence was then taken from a range of witnesses, usually including a relative or friend of the deceased, anyone present at the time of death, and the "first finder" of the corpse. A medical witness might also appear before the tribunal, unremunerated prior to 1836, after which he received £1.1 for "simple" evidence or £2.2 if the coroner had ordered a postmortem examination. The coroner was in charge of selecting the medical witness; if a majority of jurors wished, however, they could nominate an alternative, a discretionary power that, according to contemporary sources, was rarely exercised.

As each witness gave testimony, the coroner took running notes, written with variable degrees of legibility and comprehensiveness. The bulk of the questioning of witnesses was done by the coroner himself, though jurors were free to interject and often did. Questions were also permitted by representatives of the deceased—a relative, a friend, or in a few cases a solicitor retained by an interested party. At the conclusion of the testimony, the coroner summed up the evidence for the jurors, who then either retired to deliberate or (more frequently) conferred briefly on the spot and agreed upon a verdict. This was in turn recorded on the inquisition form and signed by the jurors and the coroner.

Described in these terms, the office, ancient though it might be, hardly seems to merit the veneration of a veteran radical publicist like William Cobbett, who in 1833 embraced the inquest as "a great favourite."[11] Yet its claim to popular allegiance, Cobbett and others argued, was based on more than mere time. The vestigial signs of ancestral intent still legible on the early nineteenth-century inquest confirmed, in their view, the inquest's standing as a bulwark of English liberties.[12] As an elected official, the modern coroner could be figured as a holdover from an earlier era of civic independence and as a crucial check on the growing influence of a centralizing administrative apparatus. The role accorded to the inquest jury stood out as another of the inquest's obvious popular features. Verdicts were the responsibility of a local jury, with coroners formally standing in relation to their juries much as did other judges—as advisors in matters of evidence and law but unable in principle to dictate the final verdict. Contrary to the tendency in other tribunals toward control of the jury, moreover, jurors were encouraged to (and frequently did) question witnesses. Juries were also entitled to append riders to their verdicts as a means of reflecting extralegal, communal judgment regarding the conduct of individuals or corporate entities. Though not a part of the formal verdicts, these riders could express praise or blame, make recommendations for reform, and even comment on the proper scope and operations of the inquest itself.[13]

The activist capacities attributed to the jury fed into a long-standing view of juries as popular tribunals of mitigation, competent to resolve specific disjunctures between law and social or political norms. This positioned jurors as (potentially) powerful representatives of popular resistance to ill-exercised authority at any given historical moment.[14] But as this function was a contested one, it also fed into a polarized view of where juries fit on the scale of social and political respectability. Like its counterparts in other branches of English law, the inquest jury was often derided (as low, base, ignorant, besotted, venal) and sometimes revered (as steadfastly independent and vigilant of civic liberties). In short, an inquest jury could equally be figured as a pristine oracle of public opinion, an example of the limits of public opinion and participation in matters of law and governance, a quaint irrelevance, or an impediment to accurate inquiry.[15]

Another important matter for dispute centered on the inquest as an "open" tribunal. Inquests were often described as being subject to the traditional principle of open "public justice," though this was not a statutory requirement.[16] Commentators regarded the inquest as procedurally looser than other forms of judicial inquiry, often referring to it (both in praise and in scorn) as a "rough and ready" proceeding. This rested in part on the proposition that the inquest's purpose was an open-ended one of information gathering rather than a formally structured proceeding against a named suspect.[17] The lack of an accusatorial framework made for a more amorphous standard for instigating inquiry; an inquest might legitimately address itself to "suspicion" or "rumor" that a death had occurred in some way out of the ordinary course of nature without needing to specify the subject of such suspicion. The dubious legal standing of the material generated at inquests, largely unfettered by the development of rules of evidence affecting other courts, added to its air of indeterminacy. "According to the best opinions," Sir John Jervis's authoritative nineteenth-century treatise on inquest law acknowledged, "the Coroner's inquisition is in no case conclusive."[18]

But there was another connotation of openness, one more loosely associated with the physical and performative elements of inquest practice. Until the end of the century, inquests were held primarily in public buildings (most often in pubs) and in the district where the body lay, as the place where the body was found determined which coroner had jurisdiction over the case.[19] The inquest's relationship to the physical body involved the proceedings in a further enactment of "openness" peculiar to it: to be a valid inquiry, the jury and coroner had to "view" the body, thus unambiguously locating a rite of access at the heart of the inquest. Inquests were locally and swiftly convened, generally proceeding within days of an inquirable death and paced in no small part by the sure onset of putrefaction in its primary evidentiary and symbolic referent (the dead body). This meant that inquests were seen as episodic, contingent affairs, dictated to by circumstance rather than legal form. The inquest's peculiar liminality was recognized, and often celebrated, well into the nineteenth century, as an 1883 *Spectator* editorial amply demonstrates: "The Coroner's Court is full of gossip, but it is sifted gossip, and it is much better that gossip should be sifted there than it should float around unsifted, to poison a whole countryside . . . The popular content may not be a result which would be worth the time of a High Court, and a heavy Bar, and a cloud of witnesses; but it is worth the time of that very irregular, but very efficient, little tribunal, a Coroner's Court, which constantly, and with its gossipy way, sweeps away a rumour which otherwise would be miasmatic."[20]

Elections, juries, and a semblance of publicness, then, were the constitutive materials out of which a popular inquest might be built, but in themselves they were open-ended characterizations. After all, elections and juries were longstanding features of English legal and political theory as well as practice, bearing no necessary or direct relationship to conceptions of popular sovereignty, and notions of "openness" were equally underdetermined. To serve as characteristic signs of the inquest's popular essence, therefore, they, like all argument from precedent, had to be placed within a framework of historical interpretation connecting contemporary concerns with selected signs of ancient practice.[21] This is what happened with the inquest during the period under examination: the inquest's association with "English liberties" figured broadly as an active and constitutive feature of the discussions about the inquest's past, present, and future, serving as a "framing discourse" for medical and nonmedical commentators alike.

The politics of the modern inquest did not simply frustrate the project of medical reform. Instead, conceptualizing a place for medical expertise at inquests was in significant ways a political project:

an expert-oriented inquest entailed a new set of arrangements, agents, and relationships of representation and representability, which amounted to setting up a new account of the politics of knowledge production. It follows that efforts to cast the inquest as a medically sensitive tribunal were often directed in ways not obviously recognizable as medical. Not only was the rhetoric of reform cast in a broader idiom, but also the unsettled relation between public and esoteric knowledge was a pivotal factor in generating a legitimizing framework for medical priority at inquests. The work of recasting the inquest took place in an interstitial space in which a dogmatic insistence on a privileged place for scientific inquiry was rarely left untempered by a recognition of the at least symbolic power of the inquest's popular features.

II

"Experts" are key figures in the history and historiography of the modern state, in large part because of a convergence between their model of knowledge making and the increasing stress placed on disinterestedness as the legitimating grounds for governmental action. Expert authority operates on the basis of detachment, secured through the carving out of fields of investigation, interpretation, and intervention that are deemed to require their own distinct (and contextually discontinuous) form of competence. Variously described in recent analyses as a dynamic of "disembedding," as a shift from embodied to impersonal and institutional mechanisms for the cultural distribution of trust, and as the effect of a peculiarly modern and "abstracting" way of knowing and representing "the social," this carving-out process entails both a detachment of public standards of action and judgment in the areas designated "for the experts" and the simultaneous formation of an insular code of practice by which such experts operate.[22]

With these ideas in mind, it is not surprising that the theme of loss and alienation often underlies histories of expertise. In such accounts the public loses not only an acknowledgment of its competence in matters heretofore considered within its sphere of legitimate activity, but also often its actual capacities. Expertise goes hand in hand with and is predicated upon a real subjective transformation.

The precise character of this loss, of course, depends upon the process under analysis. In political terms, the age of the expert tends to be associated with the rise of bureaucratic and centralized administrative structures. This shift, referred to in the context of British historiography as the "revolution in government,"[23] was, of course, viewed with some concern by contemporary commentators, not least among them liberal theorists whose model of politics and society provided the broad framework within which the "revolution" was itself taking place. Expert governance posed something of a paradox for them. In John Stuart Mill's estimation, one of the great dilemmas of representative government was how to strike a balance between "government by trained officials" and "participatory government," how, in his words, "to secure, as far as they can be made compatible, the great advantage of the conduct of affairs by skilled persons, bred to it as an intellectual profession, along with that of a general control vested in, and seriously exercised by, bodies representative of the entire people."[24] For Mill the benefit of expert governance, particularly in an era of mass democracy, was that it could draw upon advanced, universalizing knowledge in the service of public well-being and, ultimately, public education. Its shortcomings, however, lay in its tendency to stifle the very instruments of civic education—the local, participatory institutions in which an active, informed, and morally elevated citizenry was forged.

Mill attempted to square these two conflicting ideals by reference to "publicity," by positing a stable (though avowedly problematic) split between the superintending functions of a watchful public on the one hand and the functions of an expert-driven executive on the other. A similar set of anxieties evinced by classical liberalism in the age of mass democracy and their (tentative) resolution by appeal to the powers of publicity have found an echo in more recent and influential analyses of political and social "modernity." Like Mill, Jürgen Habermas considers the capacity for a shared and interventionist civic mode of analysis and judgment—his "public sphere"—of paramount importance in a liberal polity. He also sees it as a fragile historical artifact, one threatened by a series of fragmenting pressures, of which the proliferating insular discourses of expertise were but one component. The esoteric framing of matters of public import vitiates a formerly active public sphere and transforms the function of publicity from an instrument of "critical authority" to a "staged display" performed for members of an essentially passive polity.[25] Just as the theoretical functions of publicity are of signal importance in conceptualizing the relationship between experts and their public, so, too, must they occupy a central place in an account of the politics of the modern inquest. But, unlike its role in the liberal schema, in my analysis publicity features not as an exogenous factor that seems to hold out the promise of resolving theoretical and practical tensions in the modern inquest, but rather as an element in its own right—at once fundamental and protean—in the debates over how the inquest could and should be understood.

The degeneration of public capacities that figures so prominently in the Habermasian analysis occupies a different, but comparably important, place in the vast historical literature on professionalization. Accounts of medical professionalization, to take the most pertinent example for a history of the modern inquest, commonly acknowledge the interaction between internal claims of expertise and the wider public context in which such claims operated. Casting medicine as a distinctive domain of expertise depended not merely on knowledge claims, but also on the ability to insulate medical practice itself from broader public standards of evaluation and judgment. In the words of Jeanne Peterson, a leading historian of English medical professionalization, "Authority came to the experts as the public was increasingly closed off from knowledge of their work."[26]

Medicalization, a term most often used to denote the progressive expropriation of health from the public sphere and its relocation in an exclusive professional domain, represents an important variation on this theme: through the process of medicalization, preexisting social understandings of and responses to such basic human experiences as pain, illness, and death are seen to be displaced by arrangements that both produce and legitimate a narrower set of expert interventions. These experiences come to be regarded in medicalized society as resting outside the public's competence, a radical redistribution of knowledge and practice the effect of which, in the trenchant analysis of Ivan Illich, is to render the public "passive."[27]

Finally, and most obviously relevant for an account of an institution dedicated to managing death, important examples of the prodigious literature on "the modern way of death" pay close attention to the double phenomenon of medical expropriation and an alienated, disabled public. The sociological investigations of Geoffrey Gorer and the monumental historical narratives of Philippe Ariès, to take two highly influential accounts, regard death in earlier epochs as a profoundly public event, one that has become, if not desocialized, at least transformed by the social power of medical expertise.[28] In Ariès's view, the modern "invisible death"—isolated in hospitals, technically managed, and physically anathematized—reflects a historical process through which the public has gradually ceded any meaningful relationship to death as a physical or a cultural event.[29]

Despite profound methodological differences, historians of medicine generally agree that modern death has become increasingly contained within the institutional and cultural apparatuses of scientific medicine. The foundation of the medical profession's empire over death lies in a twofold shift in the relationship between disease and the body. First, medical theories of death since the mid–eighteenth century have progressively restricted death to physiology, locating and thereby limiting it to increasingly specific sites in the body.[30] Coextensive with the increased specificity of a bodily "topography" of disease and death was the displacement of an ontology of health and disease by one operating on the differential principle of "normality," in which the pathological state was no longer conceived of as wholly other to the healthy state, but instead differed only in degree.[31] These conceptual shifts had important implications. Explaining death by physiology restricted the field of vision in matters of health and death to an insular, desocialized body, while the new conceptual framework of pathology alienated bodily processes from the subject's comprehension, experience, and ultimate control, replacing them with normative standards developed from within and therefore possessed by the "science of life."

The participation of medical and scientific expertise in the formation of the interlocking political, social, and cultural changes classed under the rubric of modernity and constitutive of a fundamentally redistributed field of knowledge and practice is thus a central theme in contemporary historiography. It is also at the core of my own study, but it appears here in a more problematic, if not subversive, guise. In the preceding and necessarily compressed survey of the scholarly literature on modernization, professionalization, medicalization, and the like, I have suggested that, even when they pay attention to the broader context in which claims to scientific authority are made, these accounts typically cast the relationship in binary terms. Public models of knowledge, participation, and judgment are taken as external to that of science, as obstacles to the full articulation of an internally generated and essentially self-sufficient scientific alternative. If the legitimacy of science as a public discourse is contested in these narratives, such contests are not seen to be located "within" science. Instead, they unfold along oppositional lines, animated by the dynamics of resistance and imposition, in which the content of each position is fundamentally stable and internally consistent. When science eventually (and often with seeming inevitability) "wins," it does so on its own terms and is largely unaffected by the traces of its agonistic ascension. The objective of the present work, by contrast, is to push at the limits of this scenario by demonstrating the degree to which the tension between scientific and public models of knowledge is played out within the very sphere designated as "science."[32]

III

I have arranged the material gathered for this study to emphasize the multiple lines of tension between persistence and reform that mark the making of the modern inquest. In doing so I have adopted a thematically recursive rather than a chronologically linear expository framework. Chapter 1 opens with the (unsuccessful) electoral efforts of the crusading medical reformer Thomas Wakley to win a position as a Middlesex county coroner in 1830. In making his case for a medical coronership, Wakley urged the compatibility of expert and public knowledge: medicine, acting in concert with tribunals of public opinion, could promote social harmony and stability. The discussion then considers the broad historical and theoretical grounds upon which this vision of the inquest as an exemplary tribunal of public liberties itself rested and suggests that its "ancient" lineage was in important respects a contemporary political artifact.

Chapter 2 takes up the question of how a medically oriented model of inquiry was meant to interact with the demands of an actively inquiring public. It examines the possibilities and limitations of this hybrid version of the coronership on its own terms and as it related to other contemporaneous frameworks for making sense of public mortality. Of these, none was more central to the elaboration of a medicalized inquest than the emergent and increasingly demanding system of civil death registration. Public health–minded leaders in the statistical movement insisted that their knowledge depended on a complete and correct record of causes of death and regarded inquests as a dangerously loose supplement to the certification regime. The cause of vital statistics thus constituted a powerful rationale for the scientific streamlining of inquest procedures. At the same time, however, leading members of the medical community regarded the very "unscientific" features of public inquiry as a potentially useful professionalizing tool. By holding inquests in cases that were not attended by registered medical practitioners, they argued, the public might be taught, or at the very least compelled, to seek out sound medical advice.

Chapters 3 and 4 analyze the various attempts of medical reformers to restructure the relationship between the public and the dead body so as to carve out a more or less hermetic space for scientific expertise at inquests. Chapter 3 examines the logic underlying efforts to transform the physical context and operational dynamics of inquiry. At the center of these visions of a "purified" inquest lay the dead body: discussions about where it was to be placed, how it was to be treated, and what relationship it should have with its public interlocutors were the manifest content of the reformist discourse and served as an indicator of what was thought to be at stake. What emerges from this analysis is a constrained exercise in constructing a separate space for science: the physical recasting of the inquest sought not so much to abrogate the principle of publicity as, by stripping from "participation" forms of access that hindered a scientific contemplation of the body, to redefine in practice the terms of purposeful access itself.

Chapter 4 remains focused on the project of framing the body as a scientific object at inquests, analyzing the work of differentiating between medical interrogators of the corpse. It also further specifies the simple model of public access at inquests by constituting as the proper bearer of medical evidence someone disconnected from the public. Calls for the development of a battery of specially trained, state-salaried pathologists to conduct postmortem investigations to the exclusion of general practitioners combined arguments based on distinctive expertise with those linking expertise to independence. "Scientific democracy" as represented by the expert pathologist held out the promise of a universalist discourse of death that transcended the individual case. By directing the problem of medical evidence inside the profession itself, chapter 4 helps to further break down the monolithic category of medicalization while sharpening the contemporary discussion of the nature of medical expertise: What language should it speak? What modes of demonstration were entailed? What were the distinctive markers of authoritative knowledge of the dead body? What degree of sympathy was appropriate to an inquiry framed for public assurance?

Chapter 5 turns the conceptual framework of medically oriented reform upside down by considering the case of deaths produced within the sphere of recognized medical practice. Concentrating on the issue of deaths under anesthesia, the discussion follows the logic of medical attempts to construct a limited procedure of public scrutiny in these cases. At anesthetic inquests, medicine served not as an investigative adjunct, but rather as the object of investigation and judgment. This reversal brought the ongoing tensions between the ideology of transparency and the needs of scientific medicine to a new level of intricacy. On the one hand, medical practice was figured as a unique form of activity not conducive to public scrutiny. The benefits of public access and public knowledge did not extend through the portals of the hospital, ultimately, because such access threatened to transform the public from a viable medical subject capable of receiving medical attention to an anxiety-ridden one, both constitutionally and emotionally unsuited for scientific intervention. At the same time it was acknowledged that an inquest might explain the "normality" of an anesthetic fatality in terms more publicly acceptable than the published results of expert panels. Maintaining a "healthily" informed public, then, was the ultimate challenge posed by this type of death.

Each chapter extracts from the ongoing discussions on inquest reform a specific set of issues that sharpened the terms of debate historically and that, from an analytical point of view, afford distinct perspectives on the central question of science's place in the public domain. The epilogue, which focuses on the making of the 1926 Coroners (Amendment) Act, is no exception. An examination of the act on its own terms could lead to the view that substantive closure had been achieved regarding the controversies of the previous century. However, when the provisions of the act are read in relation to the complex conditions of their production, they appear to signal less the triumph of medicalization over the obstacles of publicity than the reconfiguration, and thus the reconfirmation, of the tension between the two. It is my argument throughout that this tension was not disruptive—either to the inquest's internal history or to its place within the wider context of medicalization—but was instead constitutive of both.

THE GENEALOGY
OF THE POPULAR INQUEST

On 10 September 1830, two taverns in the vicinity of London's
Clerkenwell Green served as outposts for rival candidates in an election that the leading radical politician Henry Hunt professed "more
important than that [for] members of Parliament."[1] The Crown Tavern was festooned with "handsome red banners" recommending
William Baker, a Limehouse solicitor and vestry clerk for Saint Anne's
parish, to the attention of the Middlesex freeholder electorate. At
the nearby Northumberland Arms, rival blue notices touted Baker's
better-known opponent, the surgeon, journalist, and redoubtable
public figure Thomas Wakley. Shortly after 9:00 A.M., Sheriff Sir
William Henry Richardson, flanked by county officials ranged in
front of the Sessions House, opened the contest by congratulating
the crowd on being assembled in the exercise of "one of those invaluable privileges with which the laws and constitution invested the
people of this country."[2]

In so doing, Richardson opened a ten-day contest remarkable in
many respects; for the modern reader perhaps in no way more so
than that the high office to which Hunt referred was that of county
coroner and the privilege Richardson exhorted his audience to exercise that of choosing a replacement for the late incumbent John Un-
win. Both candidates' list of supporters confirmed the importance
of the choice ascribed by Richardson and Hunt: Baker called local
dignitaries to his side, led by the recently defeated parliamentarian
Samuel C. Whitbread; Wakley's cause drew the active support not
only of Hunt but also of the most influential of his radical colleagues,
including William Cobbett, Joseph Hume, and Francis Place.

The presence of these political luminaries and the tributes they
paid to the inquest as an institutional bulwark of English liberties

raise immediate questions about the stakes attributed to this contest. On what grounds did the participants agree that the coroner's inquest was an important constitutional tribunal, and why should it be associated with radical politics? A further set of questions is suggested by the way these invocations of ancient liberties mapped onto the two candidates' professional affiliations. Baker's credentials as a lawyer well connected in parochial administration made him the archetypal candidate for the coronership of his day, whereas Wakley's bid to capture the office on the strength of his medical credentials was a decidedly novel one. From a historical perspective, however, Wakley's stance might well seem the more recognizable of the two, with his campaign representing one instance among many of the steady expropriation by professionalized medicine of hitherto lay spheres of competence and judgment.

Yet Wakley's case oversteps the conventional limits of such a "medicalization" analysis by encompassing the very discourse of legitimation commonly taken as antithetical to the claims of scientific expertise, that of participatory democratic politics. By making himself available to the Middlesex freeholders, Wakley was offering an alliance between medical science and popular politics, positioning the coronership as a fulcrum of reform and the inquiry into the causes of accidental, suspicious, or otherwise unnatural death as an extension of radical politics. Far from unfolding neatly within a space cleared by these ostensibly polarized historical trajectories, the election provided a pretext for articulating a complex set of connections between the narrative of science's inexorable march onto the public stage, on the one hand, and of the stalwart defense of the rights of freeborn Englishmen, on the other.

At the hustings Wakley recommended himself to the electorate with four slogans that purported to fuse his hybrid message: "Wakley and Medical Reform," "Wakley and the Sovereignty of the People," "Reason and Science against Ignorance and Prejudice," and "Wakley and an Open Court."[3] The first two drew attention to Wakley's credentials as a public man: Wakley the outspoken champion of wholesale reform in the medical and the political realms; Wakley the crusading editor of both the *Lancet* and the (short-lived) *Ballot,* who conceived of journalism as properly concerned with "opening"

the closed, corrupt worlds of medicine and politics. Wakley the political spokesman advocated the full range of radical causes; he
frequently attended and occasionally chaired meetings of organizations like the National Union of Working Classes, and he took such
opportunities to preach the virtues of a government open to the
scrutiny and influence of the people. Wakley the medical journalist
used the *Lancet* as a tool for prizing open the closed, patronage-
bound world of the London medical elite. In the words of one ad-
mirer, the pre-*Lancet* medical world was the undisputed province
of "men who obtained their offices by intrigue, with whom all depends upon influence and favour, and nothing upon intelligence,
knowledge, or capability." But, he continued enthusiastically, "THE
LANCET appeared—threw open its pages for the publication of every
act of injustice or oppression that was committed, and afforded that
best shield against the tyranny of the powerful—publicity."[4]

To realize its public mission, however, the inquest required the
assistance of medical science. Hence Wakley's third slogan: science,
in his estimation, was a tool of exposure, an extension of the
searching rational eye. Wakley's cause, as one of his supporters
declared at the hustings, was that of "enlightenment and humanity," representing progressive knowledge in both its scientific and
political applications.[5] "Wakley and an Open Court" further cap-
tured this promised complementarity of science and politics. Though
methodologically distinct from other paths to truth, science flour-
ished under the same conditions of openness required, from the
radical perspective, of all inquiry. Legitimate, uncorrupted medicine had a different relationship to truth than did lay professions
like the law. Where a lawyer was "fettered by legal sophistry and
precedents," a doctor cared only for unencumbered and socially unmediated evidence.[6] While a lawyer's trade was irreducibly social,
embedded in contingent vested interests, a medical man was connected to the social realm only at a removed level, in that he might
read out of his object of professional investigation signs of dysfunction, abuse, or corruption (but not himself be their cause).[7] This
also meant that Wakley's form of medically led inquiry was better
equipped to honor the traditions of open justice than was his legal
adversary: "His Court should be at all times open to the Press; in

the fullest publicity would be found that open and severe scrutiny, without which substantial justice would not appear to the public to be done. He was resolved that everything he did should be in the open face of day." A lawyer coroner, by contrast, could not afford the price of publicity. "If [Baker] did only keep an open Court for three months," Hunt declared from the hustings, "it would suffice to show his incompetency, and he would be discharged."[8]

Wakley and his allies claimed the coronership on the strength of the investigative method and professional epistemology at their disposal and its potential for furthering the cause of liberty. But their promise of an objective medical coroner looking only to the people's interests struck others as far from self-evident. For Baker's supporters a medical coroner contradicted the very essence of the inquest's constitutional standing, being a harbinger not of enlightened liberty, but of faction and secrecy. A medical coroner would introduce a "pernicious" form of professional prejudice. Medicine, in George Young's estimation, could lay no legitimate claim to universalist truth. Instead, it occupied the decidedly more mundane space of self-interested sectarian conflict: "There is no unerring standard on medical questions," Young insisted. "The judge will bring to that bench where strict impartiality should preside, his own dogmas and prejudices, and prepossessions. He will draw the attention of the jury from the plain and straightforward investigation of facts, into the labyrinths of his own scientific inquiries."[9] Medical science, then, was itself unfit to sit in objective judgment. There was no such thing as Medical Knowledge, only knowledge*s* emanating from located and interested sources. In the hands of a medical man, the coronership would become a public pulpit for preaching the virtues of a particular medical regime cloaked in the language of scientific objectivity. Instead of serving the public interest, the inquest would render the public an object of medical professionalizing strategies, medical propaganda, and medical curiosity.

At a deeper level, Young maintained, a coroner dispensing expert medical advice to the jury from the bench threatened the essence of the inquest's constitutional standing. Whenever Wakley protested respect for the independence of the jury, Young insisted, he only revealed the incoherence of his candidacy. If as coroner he respected the jury, "his medical knowledge is useless." If, on the other hand, Wakley were to lead the jury through the evidence on the basis of his own medical knowledge, Wakley's coronership would become worse than useless: the verdict would "be returned, not, as the jury have sworn, 'according to the evidence,' but according to the opinion of the judge." Wakley's candidacy ultimately represented nothing less than "an insolent attempt to unite in one the offices of judge, witness, and jury, and virtually to abolish in this portion of our judicial institutions one of the most cherished bulwarks of rational liberty."[10]

This question about the proper relationship between the needs of medicine and the needs of the public remained central to debates about the nature and function of the inquest for a full century after the Wakley-Baker election, albeit in differing forms and degrees of intensity. The modern inquest was in this sense forged out of a powerful, persistent, and often unstable confluence of the ostensibly divergent historical narratives of participatory democratic politics and medical expertise. These two themes and their relationship form the basis of the opening chapters of this book; the present chapter provides an account of the complex genealogy of the popular inquest on display at the 1830 Middlesex election, while chapter 2 examines its fusion with medical arguments for an alliance between the inquest and science.

I

In mapping out the foundations of the inquest as a feature of English liberties, this chapter takes as its point of departure the partial reading of the inquest's remit publicized by its nineteenth-century populizing proponents. In their view, the inquest was a traditional bulwark by which the people could resist the ever-looming threat of abuse of authority. Here, one feature of the ancient inquest stood out over all others, namely, the duty to hold an inquest into the death of every prisoner. In seeking to ground the inquest as a singularly necessary component of the modern administrative state, accordingly, reformers were apt to begin with the precedent of prison inquiry.[11]

Henry Hunt's presence at the Middlesex hustings was both an

illustration and an amplification of this principle. A veritable icon of radical politics, Hunt's own personal and political history intersected with the major landmarks of antiauthoritarian struggle out of which the popular inquest was being forged. These were credentials that he was not reluctant to advertise. In a speech delivered before a gathering of some four hundred Wakley supporters at the famed Crown and Anchor tavern, for example, Hunt recalled the actions of two past coroners, actions that presented mirror opposites to Wakley's "determination to resist power in high stations, when that power was in his opinion exercised to the oppression of others." The first was the conduct of the Oldham coroner in a case "held about eleven years ago, which was adjourned from day to day, from which the public were excluded by direction of the coroner, and which was finally quashed through an informality discovered in its proceeding, and thus they got rid of any verdict." The second involved the Ilchester coroner, whose collusion with an "abusive" local prison regime had constituted a fatal check on the principle of independent public inquiry. For his audience, these were unmistakable references to connections between the inquest and the foundational event of modern radicalism, the 1819 Peterloo massacre and its aftermath.[12] To appreciate the value of the coronership, Hunt was saying, one needed to know one's history—the interlocked histories of English popular institutions and of the freeborn Englishman's extended struggle to recapture his rightful liberty.

The link between prison death and inquests was not an invention of nineteenth-century radicals. The principle of prison inquests had been confirmed by successive generations of legal commentators since at least the late thirteenth-century declaration in *Britton* that "if any person die in prison, our pleasure is, that the coroner go and view the body and take a true inquest of his death, in what way it has happened."[13] *Britton*'s basic claim was confirmed by generations of subsequent commentators, such as Coke, Hale, and Blackstone, and by the early nineteenth century the pedigree of the prison inquest had become a well-codified article of legal faith.

However, the suggestion made by modern reformers that the inquest stood on the side of an informed yet vulnerable public against abuse by state authority attributed a rationale to prison inquests that would have sounded odd to *Britton*'s readers and, indeed, to legal scholars for centuries to come. Prisons, of course, had never been considered benign institutions, and from at least the mid–fifteenth century periodic efforts were made to regulate them, efforts often spurred by scandalous cases involving destitution and death.[14] Yet the inquest as a buffer against despotic abuse was not a noticeable feature of the literature on prison reform. Though the Leveller and legal scholar John Lilburne appealed from prison to the freemen of London to oppose the Crown's practice of committing their fellow citizens "to the severall murthering-houses (stiled Prisons) in this Kingdome, abounding in cruelty, murther, and oppression," he did not include an account of any recognized procedures for inquiry into such murderous practices. Inquests were similarly absent when, at the end of the century, the incarcerated radical James Whiston denounced prisons as "slaughter-houses," claiming that by firsthand experience he had seen "that through the cruelty of the *Prison-*keepers, such great numbers of poor People have been stript to their naked Skin, and when all was gone, have been Suffocated in *Holes* and *Dungeons,* to the loss of many of their Lives." Neither did inquests figure in the writings of leading eighteenth-century prison reformers like Jonas Hanway and John Howard: Howard spoke of inspection as a check on abuse, but never the inquest as a valuable (if belated) form of "inspection."[15]

Even when early modern tracts explicitly linked prison death to inquests, these hearings were not depicted as occasions for public exposure. In accounts alleging willful murder in prison, like those by Robert Ferguson (1684) and Lawrence Braddon (1692) concerning the notorious death of the Earl of Exeter in the Tower of London, inquests figured principally as exercises in obfuscation.[16] The London publisher Jacob Ilive, imprisoned in Clerkenwell House of Correction for libel in the 1750s, dismissed inquests held in that institution as "rather Matters of Ceremony . . . than of Utility."[17] Indeed, if a theory of sovereign wrong was embedded in the early prison inquest, it was not that of the public against the power of the state but that of Crown prerogative injured. An example of this framing can be found in the diary of the Quaker radical Thomas Ellwood, who witnessed a prison inquest while incarcerated at

Newgate. As soon as the jury entered the room in which the deceased had been held, Ellwood records, the foreman exclaimed: "Lord bless me, what a Sight is here! I did not think there had been so much Cruelty in the Hearts of *English* Men, to use *English* Men in this manner! . . . If it please God to lengthen my Life till to Morrow, I will find means to let the *King* know how his Subjects are dealt with."[18] It is the king who is invoked as the wronged party and the authority to which an appeal must be made, whereas the "public" is noticeably absent from this and other like accounts.[19]

This makes sense when one considers the place of the inquest in the broader context of law and sovereignty in the medieval and early modern periods. The origins of the inquest lie not principally in an ancient version of a participatory social order underwritten by a regime of publicity, but rather in the medieval conjuncture of the Crown's judicial and fiscal interests. The office of the coroner was created in the late twelfth century to function primarily as a local agent securing the king's interest in the revenues deriving from the administration of justice. The name itself derived from the coroner's obligation to record and thus "keep" all Crown pleas between visitations of the general eyre: "The function implied by their title," Pollock and Maitland explained, "is that of keeping (*custodire*) as distinguished from that of holding (*tenere*) the pleas of the crown; they are not to hear and determine causes, but to keep record of all that goes on in the county and concerns the administration of criminal justice, and more particularly must they guard the revenues which will come to the king if such justice be duly done."[20]

Such pleas were kept, in the first instance, as a check against the encroachments of local officers such as sheriffs upon Crown prerogative. Coroners protected the king's interest by enrolling a host of forfeitures and other penalties owed to the Crown as compensation for malfeasances on the part of individuals and communities of the realm. Deaths the causes and circumstances of which had monetary consequences for individuals and communities were one such class of pleas. Accidental deaths caused by a moving, inanimate object, for example, entailed a forfeiture to the Crown.[21] Other actions (or failures to act in a prescribed manner) in cases of death could similarly result in levies: the failure to raise a proper hue and cry upon the discovery of a body or the burial of a body before its inspection by the coroner, for example.[22]

Inquests into deaths, though undoubtedly the most frequent duty exercised by the medieval coroner, formed only one subset of the several Crown pleas that he was obliged to manage. Inquiries into treasure trove, wreck of the sea, and the finding of royal fish (sturgeon and whales); abjurations of the realm (by which felons confessed to their crimes and were exiled); appeals and outlawries; and rapes also fell under his jurisdiction.[23] The eminent medieval jurist Henri de Bracton, in the first comprehensive account of coroners' functions (recorded some fifty years after the establishment of the office), offers a template for this seemingly inchoate grouping of inquiries. Bracton discusses the duties of coroners in his chapter on treasonous attempts "against the king and his dignity and his crown." These included denying the king the resources that were duly his (e.g., the "fraudulent hiding of treasure-trove") and the slaying of one of his subjects, which "touches the king himself, whose peace is infringed."[24] It is in the context of these affronts to sovereignty that Bracton enumerates the coroner's functions, beginning with those relating to homicide and followed by brief paragraphs on his duties as to treasure trove, the rape of virgins, and peace and blows (wounding).

Sir Thomas Smith's classic sixteenth-century account of the contemporary English legal world provides a rationale for the expositional context that Bracton and others following him selected for their discussion of the inquest: Because "the death of everie subject by violence is accounted to touch the crowne of the Prince, and to be a detriment unto it, the Prince account[s] that his strength, power and crowne doth stande and consist in the force of his people, and the maintenaunce of them in securitie and peace."[25] The wording of inquisitions into death further underscores this conflation of sovereignty and accountancy. Seventeenth-century inquest jurors were advised that they convened "to the end that the King and his immediate Officers may be truly certified how and by what meanes he hath lost his subject." Coroners, for their part, were bound by their oath of office to uphold the Crown's complex interest in a subject's death: "You shall diligently and truly doe and accomplish all and every thing and things appertaining to your office, after the best of your cunning, wit, and power, both for the Kings profit, and the good of the inhabitants within the said county, taking such fees as you ought to take by the lawes and statutes of this realme, and not otherwise."[26]

Sovereign dignity and profit, at stake in every inquest, was implicated in the particular instance of prison death by virtue of the Crown's prerogatives in imprisonment and the infliction of bodily punishment. As a general principle, since prisons were coextensive with the person of the king, any wrongdoing on the part of the franchise holder constituted an affront to his dominion.[27] Prison deaths, furthermore, represented a breach of sovereign interest with a unique set of consequences. As John Langbein observed in his work on torture, medieval incarceration served a well-recognized coercive function, "designed to compel someone to take some other procedural step, characteristically the payment of a crown debt or a civil judgment debt."[28] In this sense prison deaths attributable to abuse or mismanagement could be counted as direct losses to the fiscal well-being of the Crown, requiring compensation of some sort.[29]

The coroner's original mandate with respect to prison deaths thus reflected the contemporary imbrication of justice, finance, and sovereignty. Medieval and early modern conceptions of the king's responsibilities for public justice, to be sure, did not resolve themselves into simple matters of fiscal accountancy. Dispensing justice for its subjects was a fundamental duty of the Crown, one that by the early fourteenth century formed part of the coronation oath itself.[30] Still, the notion of a kingship bounded by an obligation to the liberties of his subjects, even at its most robust articulation, was a long way from an active, self-informing, and self-protective "public." Prison inquests as tokens of the transhistorical tenderness of English common law were artifacts of a much later political reconfiguration in which the Crown yielded to "the public" as the guarantor of order and the injured party whenever that order was breached.

Nor did the inquest move center stage as a consequence of the long-term shifts in theoretical and practical notions of political sovereignty through which, by the end of the eighteenth century, "the public" came to serve as the key referent for a radical politics. This was true even when prison abuse itself stood out as an explicit emblem of the battle against government despotism in the name of the people, as the high profile campaigns for prison reform waged by the Westminster radical Francis Burdett amply demonstrate. When, in July 1800, Burdett moved in Parliament for a commission of investigation into three deaths at Cold Bath Fields (the recently constructed Middlesex House of Correction), he explicitly distanced himself from the mechanism of prison inquests. The deaths of the prisoners, he asserted, had been inquired into by a "packed jury," and thus the inquests could not be looked to for the truth. Burdett further observed—in a statement that reveals just how far he was from the rationale attributed to the inquest by "traditionalists" of subsequent decades—that the episodic nature of prison inquests rendered them ill-suited to the requirements of overarching reform: they might "redress the grievances of a single individual, but they could not redress those of the public."[31] It was, tellingly, a representative of the governing authorities who instead made recourse to the inquest to counter Burdett's call for parliamentary intervention: "There was a remedy provided by the law," the attorney general assured the House. "If a man died in prison, the coroner had power to sit upon his body, that the cause of his death might be ascertained."[32]

It is no great surprise that, as Burdett and parliamentary colleagues like Brougham, Romilly, and Wilberforce continued to denounce prisons as instruments of an ever-widening despotic state, they did so either without reference to the fulcrum of the inquest or by express denunciation of the inquest as an arm of "Old Corruption."[33] Burdett's strategy of disavowal was the same, moreover, when he appealed directly to "the public"—ostensibly the timeless constituency and referent of the antiauthoritarian inquest. He made prison abuse the centerpiece of his famed 1802 parliamentary contest against William Mainwaring. Mainwaring, a sitting member of Parliament (MP) for Middlesex and chairman of the county quarter sessions, had been outspoken in his defense of the regime at Cold Bath Fields prison against charges of brutality. "This is not a question whether you will support me or Mr. Mainwaring," Burdett declared in his first speech on the hustings, "but whether you will support that scandalous gaol. Whether you will support cruelty, torture, and murder by torture or whether you will support me in a fair enquiry into, and investigation of this most abominable subject."[34] But here, again, *fair enquiry* meant parliamentary, not "public," investigation.

Prison abuse also featured prominently in the early nineteenth-century writings of extraparliamentary commentators like William Cobbett. Predictably, Cobbett diagnosed the problem as a combination of innovating despotism and seeping corruption consequent upon the steady erosion of ancient practices. The local county authorities in charge of prison inspection, the magistrates and the sheriff, were fast losing any semblance of their former independence and were instead becoming constitutive features of an ascendant despotism. Accordingly, when Cobbett wrote in support of Burdett's efforts to organize a parliamentary inquiry into the 1812 deaths at Lincoln jail, he dismissed the regime of official inspection as one carried out by "mere placemen" whose activity "stands in need of investigation more than that of almost any other description of men in authority." This had not always been the case, Cobbett advised his readers: Sheriffs, for instance, were "in former times, *elected by the people.*"[35] The lost investigative efficacy of a sheriff derived from the severing of his connection to the people by an electoral covenant.

In proposing this analysis of a corrupted modern system of prison investigation, Cobbett had, of course, set the stage perfectly for invoking the one surviving mechanism of effective inquiry in such cases—that of the popularly elected coroner and his jury of local men good and true. "My readers will bear in mind," Cobbett wrote, "that Coroner's Courts are a part, and a most important part of the '*ancient institutions*' of the country." The inquest, he insisted, "has always been a great favourite of mine," a proceeding conceived "for the protection of *life and limb,* as old as the laws under which we live, and a part of the constitution of which we ought to be particularly jealous."[36] However, these panegyrics to the ancient lineage of the popular inquest were written not in 1812—when Cobbett was casting around for an independent form of inquiry into a prison death—but some twenty years later. In 1812, Cobbett had to settle for a parliamentary commission as a guarantee that the deaths at Lincoln would be made the subject of "a *real* inquiry; not a *base cheat* under the name of an inquiry." Like that of other radical com-

mentators of the day, Cobbett's romance with the supposedly timeless institutional embodiment of English liberties had yet to bloom.[37]

II

As for so many other political developments, the traumatic events
of Peterloo and the subsequent national soul-searching they provoked were key to raising the profile of the inquest as a reformist
cause. The months immediately following the bloody events of August 1819, as Dror Wahrman has suggested, saw the culmination,
on the one hand, of an unprecedented surge in the confidence accorded to publicity and public opinion as ultimate meta-political ar-
biters and, on the other hand, of a strong urge to stuff the genie of
violence back into the bottle of orderly political process. The inquest
fit perfectly into both developments, promising a means to channel
public rage into constitutional arrangements (those that were at the
heart of the constitutionalist turn in radical politics after 1815) and
to allow radical agitation to continue in distinctly counterviolent
ways.[38]

The story of Peterloo requires little substantive retelling. As a climax to a summer's campaign of mass demonstrations, Henry Hunt
led a gathering of tens of thousands at Saint Peter's Fields, Manchester, on 16 August 1819. The meeting was violently dispersed by
the combined force of government troops and Manchester yeo-
manry, leaving at least eleven mortally wounded. Among them was
John Lees, an Oldham cotton spinner and a veteran of Waterloo.
Lees's death was (and is) by far the best known of those connected
with the "Peterloo Massacre," a notoriety earned in the first instance
by Lees having survived his injuries for a full three weeks. His tenacious grip on life made the inquest into his death unlike those held
in the days immediately after the events, as it allowed time for two
solicitors connected to the radical cause to appear at the opening
day of the inquest accompanied by several coachloads of witnesses
prepared to testify against the authorities. As a consequence of the
organized case presented on behalf of Lees's family and the cause of
English liberty, the protracted inquiry into his death before the coroner for the district of Rochdale and a jury of twelve local men *probi
et legales* was played out on a national stage.

In the course of the inquest, discussion centered on "constitutional" issues; most notable were the status of the inquest as an "open" court to which "the public" had a fundamental right of access and the role of the press in transmitting its proceedings onto a national stage.[39] These connected with broad jurisprudential debates that, in the context of the political repression of the revolutionary era, had been consistently before the courts in previous years. Equally, abstracted from the specific institutional location of the coroner's inquest, principles of open justice were matters of major concern for a radical politics founded on the defense of ancient liberties.[40] The Oldham hearings definitively placed the inquest and open justice in the same frame: the coroner and his supporters argued that, in the climate of intense agitation surrounding it, the inquest was best kept out of the public eye; radicals saw in the full public hearing a chance to expose government repression to public opinion in a clear forum.

The result was more equivocal than either side might have wished. The proceedings were ultimately terminated by the High Court—at the government's behest—before a verdict could be reached. In this sense the forces of "order" won the day. But suspension also provoked powerful protest. Brougham, for one, took to the floor of the Commons to protest the sacrifice of legal principle for base political expediency: "When he saw a bare-faced collusion like that which had long been exhibited in the face of the country, when he saw a coroner questioning his own irregularity, and availing himself of his own breach of the law, to escape from the necessity of executing the law, he could not think of dignifying such a mockery of the people of England with the name of justice."[41] For the opposition press, the Lees inquest struck critics as part of a broad-scale assault on the foundations of English liberties, of which Peterloo was itself the most obvious and flagrant instance. Opponents of despotism, an *Examiner* correspondent declared, were facing a constitutional crisis of epic proportions: "As if the atrocious sabring of an unarmed inoffensive population were not sufficient of itself to mark the triumph of arbitrary principles, supported by military power, over the constitutional rights of the people, we are doomed to witness the scandalous perversion of judicial proceedings in support of that system of state policy which will either destroy this country, or it must be destroyed by the spirit of the British nation."[42] The resort to legal machination thus provided a means to invest in the popular inquest the emotions attached to the most traumatic martyrdom in the annals of radical politics. The inquest, at least in principle, emerged from Peterloo as a constitutionally sanctioned answer to the problem of reconciling the tension between appeals to public opinion and the specter of public disorder.

Small wonder, then, that the opposition press took the opportunity over the next several years to raise other cases of inquests that equally promised to expose abuse. The inquests on the two victims of the disturbances at the funeral procession of Queen Caroline in August 1821 (itself another important milestone in contemporary radicalism) provided the most arresting of these opportunities. For over a week the London opposition papers intently followed the proceedings, drawing close attention to what they perceived as attempts by "ministerial hirelings" to thwart a full and open inquiry (and making explicit reference to the Peterloo inquest in so doing). "Which of the old legislators of England," *The Times* demanded, "ever dreamed that in prosecuting the ends of public justice, where the blood of Englishmen had been shed, and where the King himself was the party most interested in ascertaining the perpetrators of the violence through which he had lost a subject—who ever dreamed, we repeat, that it lay in the breast of the executive officer to throw obstacles in the way of such an inquisition?"[43]

A series of lesser-known prison inquests—including a spate of deaths at the recently constructed national penitentiary at Millbank—also provided opportunities for criticizing the administration in the wake of Peterloo. From March to October 1823, the unsanitary conditions at Millbank and the deleterious effects of its innovative restricted diet and regime of solitary confinement upon prisoners' physical and mental health were subjects of wide press and parliamentary attention. Significantly, press accounts of a series of inquests held at the penitentiary were no longer cast in exclusively dismissive terms. At the termination of the first of these, the *Black Dwarf* supported its call for prompt inquiry into prison diet by reporting that the "coroner's inquest has declared its opinion that so many would not die in the *Millbank Penitentiary*, if a *little more food* were allowed." One month later, the *Black Dwarf* informed its readers that an inquest jury had found that another Millbank inmate had died "'in consequence of the *short allowance*, and the *bad quality of the food*, with which she was supplied before the new regulations took place in this prison.' . . . This very proper verdict," the paper remarked, "ought to fix eternal shame upon the secretary of the home department, whose *authority* must have been obtained for the regulations for starving the prisoners, and hastening their dissolution by *unwholesome food*."[44]

In a direct reversal of the position adopted by Cobbett and Burdett a decade earlier, criticisms were also laid against the parliamentary inquiry into prison conditions called in the wake of the Millbank deaths. The *Real John Bull* expressed wonder at the Select Committee's exoneration of the Millbank "charnel-house," dismissing it as "a Committee of grave old gentlemen publishing such self-evident nonsense."[45] *The Times* was only slightly less scathing in its assessment: "We confess we cannot implicitly rely upon the authority even of a Parliamentary Report for a fact which seems contrary to the evidence of our senses."[46]

But it was Henry Hunt, imprisoned in Ilchester jail for his part in the Peterloo demonstration, who both symbolically and practically epitomized the emergent radical constellation of prisons, inquests, and historic English liberties. The mere fact that it housed the nation's best-known political prisoner guaranteed Ilchester's status as a prominent institutional embodiment of post-Peterloo reaction. Hunt's voluminous prison writings from Ilchester, moreover, gave shape to the critique, especially his lengthy *Investigation at Ilchester Gaol*, which appeared in the autumn of 1821, dedicated to King George IV and signed from the "Ilchester Bastile."[47] In this critique of the despotic regime presided over by the warden, William Bridle, Hunt sought to expose (among many other grievances) the circumstances and the flawed investigation of several recent inmate deaths.

Statements by the Somerset coroner, Richard Cames, and the jail's surgeon, James Bryar, demonstrated, by Hunt's standards, a perfunctory system of investigation headed by a coroner who had no appreciation of the significance of his public responsibilities in such cases.

Cames readily described the questions at issue in prison inquests as routine and requiring little time or scrutiny: "I may venture to say that no jury ever hesitated to find a verdict for five minutes, I may say two minutes." When Hunt suggested that the jealousy of the law "requires *more time* in consequence of the person dying in prison," Cames dismissed the premise of the question. Prison inquests were no more significant than any other inquests, Cames observed laconically, and in such cases "there is a certain point to which we must come, and having come to that point, it would be perfectly useless to go beyond that."[48] Hunt pressed the coroner on the limits of his curiosity in the case of a prisoner named Ford, who had died, according to Bryar's testimony, of a rupture on the brain or near the heart. Asked by Hunt whether this was not a case for further medical examination, Cames replied: "I am not a medical man, but I have understood that persons dropping in fits and dying instantly, that their death is occasioned by an act of God of that kind; and it is not usual to have operations performed to ascertain which of the two causes, whether from the head or the heart."[49]

As Cames continued with his testimony, Hunt sought to expose the details of the Ilchester coroner's betrayal of the law's "traditional" regard for the lives of prisoners—and by extension, for the liberties of all freeborn Englishmen. He established that Cames habitually deferred to the warden's definition of established prison practice rather than seeking to determine whether such practice conformed to the dictates of common sense, common humanity, and lawful precedent;[50] he accepted medical evidence that he knew to be partial at best; he did nothing to shield inmate witnesses from possible recriminations by the prison staff, professing at one point to be "so little acquainted with the internal regulation of a prison, that I did not know whether they were prisoners or not."[51] In his responses to Hunt's interrogation, James Bryar cast more doubt on the independence and efficacy of inquests at the jail, acknowledging that he had on occasion given evidence in cases with which he had had little or no contact. Further aspects of Bryar's attendance to the prisoners' physical well-being came into question—for instance, his willingness, at Bridle's behest, to use physic as a form of punishment.[52]

Having exposed, to his satisfaction, the bankruptcy of the medical and legal procedures undergirding inquests at Ilchester jail, Hunt concluded with a dramatic enactment symbolic both of his critique and of his vision for reform. He served the coroner with the following notice: "Ilchester, June 19, 1821. Mr. Cames,—If you should be called upon to take an inquest upon my body, while I am in this prison, I hereby request that you will summon Mr. Shoreland, of Ilchester, and Mr. Davis, of Andover, surgeons, to open my body. H. Hunt."[53] By objectifying the prison inquest, by linking medical and political corruption, and by suggesting through implicit contrast a regime of efficacious public inquiry into prison death, one in which a searching medical inquiry would buttress strict adherence to the principles of open and impartial justice, Hunt helped to take prison inquests out of their previously marginalized position within the canon of radical reform. To connect Hunt's notice to his appearance with Wakley at the Middlesex hustings some ten years later, we require only a straight line.

The argument here is not that the Ilchester inquiry and the spate of politicized inquests in the wake of Peterloo definitively established the coroner's inquest as a commonly accepted institutional counterweight to government encroachments upon the life and liberties of Englishmen. Inquests in the context of prison and other possibly abusive spaces and relations could still be ignored or disparaged. As late as 1827 the reforming law journal *Jurist* could lament that the inquest "is not commonly classed among the blessings which distinguish us from surrounding nations; it is not toasted in taverns, nor called a palladium of the Constitution."[54] Nor was independent medical participation inevitably linked to the radical model of inquests as a popular bulwark against despotism. Indeed, a contributor to the *London Magazine* explicitly invoked the Lees inquest to illustrate precisely the reverse, that medical expertise needed to be developed to shield authority from the undue suspicion of an agitated populace: "Deaths, which at first were considered as the consequence of some injury, inflicted by a desperate or malicious hand, have been traced by a skilful anatomist to a very different cause . . . The professional knowledge and anatomical skill, which so often serve to detect the criminal, are no less frequently, and certainly far more gratefully instrumental, in tranquilizing the public mind." The inconclusive and contradictory medical evidence at the Lees inquest, in the writer's view, was the most recent and important example of this principle, the result of which had been to vitiate the explanatory (i.e., exonerating) value of an inquest "whose decision was awaited with the utmost anxiety by a whole empire."[55]

The point, instead, is that the complex of associations running through Peterloo and the subsequent canonization of "Saint Henry of Ilchester," drawing on notions of publicity current in both theoretical and practical articulations of the English polity, established a frame of reference that could (and would) be self-consciously mobilized in the cause of opposition politics. The resilience and flexibility of this vision of the inquest was in clear evidence when, in the charged atmosphere of the 1832 parliamentary summer session, the Commons considered a series of proposals for amending inquest law. One proposal in particular, that calling for statutory recognition of the inquest as an open forum, unambiguously reached back to the controversies of the previous decade. This measure was introduced by the veteran reformer Henry Warburton, and an analysis of the debate generated by it provides as clear a sense as any of the multivalent associations at work in the contemporary politics of the inquest.

That Warburton, a member of the extended Benthamite parliamentary circle, would take the lead in promoting the cause of a popular conception of the inquest is in itself worthy of comment. Bentham himself had no sympathy for arguments grounded in appeal to the ancient constitution, and his penchant for systematizing and rationalizing led him either to ignore the inquest altogether or to disparage its continued existence as an example of the maddening tendency in English common law to preserve "barbarous" relics in the name of this illusory ideal.[56] But at another level Bentham fits perfectly in an account of the making of the popular inquest. Among the specific targets of his prodigiously wide-ranging reformist zeal, two of Bentham's abiding concerns—the reconstitution of the prison and of the legislature—are key to understanding how this might work.

Enrollment of "the public" as a check on abuse and mismanagement was a cornerstone of Bentham's famed (and unrealized) panoptic model of prison administration: "Publicity is the effectual preservative against abuse," he wrote, adding that "under the present system, prisons are covered with an impenetrable veil; the Panopticon, on the contrary, would be, so to speak, transparent." By recourse to technical innovations like "conversation-tubes," which were to connect convicts in solitary cells to a central, public inspection room, Bentham envisioned a penitentiary house "thrown wide open to the body of the curious at large—the great *open committee* of the tribunal of the world."[57] His confidence that such a salutary committee of the curious would materialize in practice was ultimately independent of any spirit of high-mindedness imputed to the public and was instead grounded in his belief that the inner workings of the penitentiary—like any other instance of secrecy—stimulated a natural propensity to curiosity and suspicion.[58]

The serious business of rational administration, then, could be secured on what were admittedly rather dubious motivating grounds, a model that Bentham insisted applied equally to the world of modern politics. Those seeking the means of constituting a legislative body capable of efficiently commanding and sustaining the assent of the governed, Bentham advised in his "Essay on Political Tactics," would do well to start by recognizing that "the fittest law for securing the public confidence . . . is that of *publicity.*"[59] Bentham believed that publicizing the activities of lawmakers would meet with interest (and thereby lay the foundations for a system of salutary public scrutiny) because, like the prison, the inner workings of government stimulated a public desire to know: "Suspicion always attaches to mystery. It thinks it sees a crime where it beholds an affectation of secrecy."[60] Though suspicion was often legitimate, Bentham nonetheless readily admitted that the system he was articulating did not function on elevated principles: at its core, the "regime of publicity," he conceded, "is a system of *distrust.*"[61] But to those who would argue that distrust was not a desirable or even workable foundation for judicious governance, Bentham countered with an observation drawn from his inexhaustible political ontology: "This objection would have some solidity, if, when the means of judging correctly were taken from the popular tribunal, the inclination to judge could be equally taken away: but the public do judge and will always judge." Bentham thus urged openness as a means of managing an inescapable economy of suspicion.[62]

By placing terms like *suspicion, curiosity,* and *mystery* at the heart of a reformist model so universal in its application as to encompass the seemingly incommensurable worlds of the legislature and the penitentiary, Bentham not only marked out the theoretical terrain upon which a new-modeled participatory and popular inquest might be based, but also applied these terms to the very institutional setting (namely, the prison) that innovating "traditionalists" invoked in defending their view of the inquest as consistent with ancient practice. Thus, when Warburton informed the Commons that inquests held in secret bred dissatisfaction and suspicion far more damaging than any harm that might follow from openness—and punctuated his remarks by invoking "the words of a great man lately deceased, 'publicity is the soul of justice'"—he was in an important sense being true to his political and theoretical lineage.[63]

Radical members of the Commons followed Warburton's lead and extended it beyond the realm of theory by insisting on the concrete, historical connections between publicity and prison inquests. Hunt rose as a direct witness to the abuses consequent upon corrupt prison inquests: "The manner in which Coroners performed their duty wanted revision," he informed the House, recalling his incarceration at Ilchester when a fellow inmate had been "murdered in gaol by the Gaoler; the Coroner held an inquest on the body in the prison, and, by carefully excluding all witnesses, brought in a verdict of accidental death."[64] The prominent Irish MP Daniel O'Connell elaborated upon the lessons posed by Hunt's example: "Suppose that a man had died in gaol—had been murdered in gaol—and such things sometimes happened!—what security was there that the Coroner's inquiry would lead to a full and fair investigation, if the inquest could be held in secret? In all such cases, the only protection which the people could have was by the free admission of the reporters of the public Press." O'Connell then closed the circle on the radical version of the imperative of open inquests by locating it definitively on the map of live political memory: "The impunity of those who were concerned in the celebrated murders at Manchester," he declared, had been "secured by the imperfection of the law respecting the Coroner's Court."[65]

Unlike the debates in the Commons twenty years earlier, when Burdett's call for a searching parliamentary inquiry into prison deaths had met with the government's laconic assurance that inquests were the traditional and sufficient tribunals for any such investigation, now, in the immediate post-Reform political configuration, radical and government forces shared the antiauthoritarian reading of the prison inquest. For government representatives, the inquest as a tool of public watchfulness seemed a felicitous marriage of rational reform with solid constitutional principle. Lord Althorp, as chancellor and leader of the House, and the solicitor general, Sir William Hone, both rose to affirm that the vote represented not an innovation but merely a declaration "in perfect conformity with the acknowledged practice." John Campbell averred that, "for the sake of individuals and the public, it was of the last importance that there should be no appearance of secresy in the proceedings of Coroners Inquests."[66] Warburton closed the discussion with an attempt rhetorically to reconcile the grounds upon which the Whig and Radical victory rested, declaring himself pleased to hear the government's legal counsel concur that his amendment "was consistent, not only with general reason . . . but was also consistent with the old constitutional law."

For a brief moment, some notion of the prison inquest as a tool in the fight against "Old Corruption" had at least the formal blessing of the reformed house. This consensus was soon shattered. In early May 1833, the National Union of Working Classes and a coalition of other radical groups disenchanted by the post-1832 political settlement called a meeting to support a plan for organizing a national convention to press for an expanded franchise. The meeting was to be held on the thirteenth of the month at the Calthorpe Estate of Cold Bath Fields (referred to interchangeably as "Calthorpe Street Fields" and "Spa Fields") in London's Clerkenwell district. The home secretary, Lord Melbourne, declared the meeting illegal, and on the day the crowd, estimated at three thousand protesters, was met by some one thousand members of the Metropolitan Police. In the ensuing fracas one of the police officers, Robert Cully, was fatally stabbed. The inquest on Cully's body became a celebrated event, terminating with the controversial verdict of "justifiable homicide."

When the Calthorpe Street affair is mentioned by historians, it generally serves as an early milestone in the history of the London police force and its relations with an initially suspicious and at times hostile populace.[67] For the purposes of the present discussion, it can serve as an endpoint of this genealogical overview of the rise of the popular inquest: it secured the terms of a radical view of the inquest as a potential space for the ordinary citizenry to resist state encroachments, yet it also signaled to its erstwhile supporters the limits (and dangers) of this same vision.

Reaction to the affray was immediate and intense. The opposition press placed "the Ministerial Riot at Coldbath Fields"[68] squarely in the despotic tradition of Peterloo: "The Tories had their 'Massacre' on a large scale," wrote *Bell's New Weekly Messenger,* "the Whigs have now had theirs upon a small: but the life-size picture and the miniature both bear the hand of one master, who writes his name— Despotism: he paints at Spa-Fields as he did at Peterloo; and, at both places, in colours of blood."[69] This provided clear proof of the Whig abandonment of the reformist crusade, a conclusion that was further cemented by the Whig government's reaction to the Cully inquest verdict exonerating the crowd of responsibility for the death. Within days of the "justifiable homicide" finding, the government's legal experts, led by Brougham in his capacity as lord chancellor, prevailed upon the High Court to quash the verdict as "bad in law" and "contrary to the evidence," the irony of which, given Brougham's fulminations against the suspension of the Lees inquest after Peterloo, was not lost on his former Radical allies.

The Cully inquest and the legal quashing of its verdict fed directly into the process of political polarization in the wake of the 1832 Reform Act, whereby the sense of betrayal among resolute Radicals led to their increasing alienation from mainstream opinion. The starkly conflicting treatment given the Cully case in the newspaper press underscores this split and points to the inquest's newfound standing as a broadly recognized feature of contemporary political contestation. From the pages of the "respectable" press came variously weighted criticisms of the inquest verdict, out of which emerged a discourse of opposition to the popular inquest as a destabilizing innovation. *The Times*'s reaction was perhaps the most troubled of these:

> The verdict may be salutary as a lesson of temper to the police, but we cannot say that it will much increase the confidence of discriminating minds in the knowledge and sagacity of juries. Indeed, when we consider the not improbable consequences of the verdict in encouraging low ruffians to resist authority, we think that the jury, who represent themselves as having some stake in the country, may before long lament that their anxiety to repress the violence of power has made them disregard the law of the land—the surest protector of the poor against the rich no less than of the rich against the poor.[70]

Other conservative papers were less circumspect. Though protesting its reverence for the liberties of the people, the *Morning Chronicle* declared "we do not wish to see the mob exercising dominion over us; we do not wish to see the law trampled on, and Juries openly promulgating to the world that men may resist the authorities of the country." And in the view of the *Morning Post,* the danger was as clear as the suspect grounds upon which such juries acted:

> It cannot have escaped remark that of late years coroners' inquests have upon several occasions arrogated to themselves functions for which they are, from their constitution, signally unfit . . . [Jurors have] taken it into their heads to imagine that no inquiry can be too large, or too delicate, or complex for their faculties, and in cases of riot attended with bloodshed have accordingly volunteered their services to determine upon the general merits of the assemblage, and the general conduct, not merely of the public force employed to disperse or controul it, but of the government by which that force has been employed. This ridiculous assumption may gratify the shallow vanity of the individuals whom the parish constable has dignified by a summons; but we venture to say that in no single instance has it ever yet contributed to the wholesome ends of public justice, welfare, and security of the community.[71]

An equally trenchant analysis of the events on the part of the radical press, on the other hand, served to stabilize the terms of a popular conception of the inquest. For weeks its pages were filled with reports of the "riot" and its aftermath. The proceedings of scores of meetings held up and down the country in support of the Calthorpe Street jury were covered at length. It was in the wake of the Calthorpe Street affair that Cobbett enveloped the inquest in his retrospective embrace. More effusive still was Henry Hetherington, whose *Poor Man's Guardian* for weeks on end reported the accolades accorded the Calthorpe heroes—the medals struck, the banners sewn, the statues erected, the parades marched (and floated, as when a fleet of barges massed on the Thames in a waterborne procession honoring the glorious seventeen). The text of one such tribute will have to suffice:

> We cannot conclude these remarks without congratulating you on the glorious result of the inquest on Cully the police-soldier. Justice and humanity have for once, thank heaven! prevailed. The conduct of the jury, throughout the whole of their arduous duties, was noble beyond praise. Their patience—their firmness—their devotion to truth—their laborious and searching inquiries during four long days, entitle them to the admiration of their countrymen . . . Oh! How this verdict has filled the oppressed with joy! What terror and dismay has it carried into the hearts of tyrants! And how incalculable will be its effects upon the tribunals and future policy of this country![72]

By the end of the Calthorpe Street affair, the ancient tradition of the popular inquest had been well and truly discovered.

III

In the mid-1830s, the inquest as a "people's court" anchored in a tradition of watchfulness against abuse of authority had a familiar, if controversial, ring. Familiarity and controversy only deepened when, at the end of the decade, this version of the inquest won a prominent practical exponent. In 1839, Wakley (by then not only the *Lancet*'s editor but also the Radical MP for Finsbury) successfully campaigned for the coronership of Middlesex's Western district. The 1839 election was a muted version of the Wakley-Baker contest of nine years earlier, but the central arguments were recognizable: a recourse to ancient liberties, common to both Wakley and his opponents, and a pledge from Wakley to reinvigorate the inquest

"Britons Bulwark. Trial by Jury." Calthorpe Street Commemorative Banner, 1833. Courtesy, Museum of London Picture Library.

using the tools of medicine and constitutionalism to turn it into an adjunct of popular politics.

In his victory speech Wakley promised to reform the position and restore the importance that the coroner had traditionally enjoyed, pledging to do so by arguments from precedent: "In ancient times the coroner had been an important officer. He would search the early statutes to discover what his rights, authority, and duties still were." It is not surprising that Wakley's first attempt to reassert his office's time-honored remit invoked the principle of prison inquests to position the coroner as the institutional embodiment of a transparent polity. In September 1839 Wakley issued a set of new rules to local parish officers, instructing them to notify his office when (among other circumstances) "persons die in confinement," including prisons and police offices, as well as more broadly construed places of incarceration, like lunatic asylums and workhouses.[73]

Within weeks the directive was put to the test, when the first in a string of celebrated Wakley inquests was held to inquire into the death of a seventy-nine-year-old pauper, Thomas Austin, who had died on September 3 after falling into a vat of boiling water at the Hendon Union workhouse. The Austin case is most memorable among Wakley biographers for his riposte to the injudicious protest made by one of the workhouse officials that the body under inquiry had not been definitively identified as Austin's: "If this is not the body of the man who was killed in your vat, pray, Sir, how many paupers have you boiled?"[74] Its broader significance, however, lies in the way Wakley sought to position the coroner as an embattled ombudsman taking on vested local interest, a formulation evident in his summation to the jury. "Persons in authority who had been and wished to continue to be free from observation and control," he warned, "were becoming apprehensive at the prospect of having the attention of the public directed to their conduct." An inquest into the death of James Lisney at the Hendon Workhouse a year later gave Wakley the chance to expand upon this vision of the popular inquest, and once again he did so by explicit reference to prison inquests: "Now, of course," Wakley remarked at the outset of the hearing, "we are not to consider that a workhouse is a gaol; but at the same time the discipline in these places appears to be of a somewhat tense and close nature, and persons in authority in workhouses have very great powers, and powers which assimilate in some respects to the powers exercised by a gaoler."[75]

Wakley's position found support among those sharing his fierce opposition to the new Poor Law. *The Times*, whose Tory radical orientation made it a leading organ of "respectable" protest against the Whig's Poor Law commissioners and their ubiquitous secretary, Edwin Chadwick, frequently drew attention to workhouse fatalities under headings such as "Death from the New Poor Law," justifying these stark declarations by reference to the apparent obfuscation with which investigations of such cases were typically conducted. "We say, death from the New Poor Law," *The Times* explained in reference to the November 1840 death of the Saint Pancras pauper Elizabeth Friry, because "it is a plainer and less circuitous phrase than 'fever brought on by want of food and sufficient nourishment,' and is preferable, as fixing the attention at once on the primary rather than the secondary cause."[76] Here *The Times* invoked the inquest and the medical evidence presented before it as a part of the problem, drawing on a well-established tradition in the opposition press of parodying inquest form to distinguish between "official" and "humane" inquiry.

But now with Wakley as coroner in the high-profile jurisdiction of Middlesex, *The Times* could posit the inquest, invigorated by a progressive medical sensibility, as a salutary contrast to other state-based forms of inspection. In the Friry case, *The Times* reported, Wakley's commendable vigilance—"well worthy of imitation by his fraternity in general"—had already thwarted one attempt made by parochial officials to have the matter "hushed up" by registering the death as "natural." *The Times* was equally clear in its denunciation of the ensuing Poor-Law board's commission of inquiry as an attempt to smother bureaucratically the judgment of a representative public institution, condemning "closed door" investigations "into matters already investigated before the proper legal tribunals, as at once casting an unwarrantable slight upon a jurisdiction more constitutional and more impartial than that by which it is sought to be superseded, and tending to create suspicions that the new inquiry is designed more to disguise than to elucidate the truth."[77]

Over the course of the following decade, Wakley presided over numerous cases such as these, thereby drawing attention to both the anti–Poor Law movement and the radical vision of the inquest. Workhouses were not the only object of his inquisitorial scrutiny; Wakley took on other supposed bastions of lingering despotism, none with more explosive effect than a case focusing on discipline in the British Army. The inquest into the death of Corporal John Frederick White at the Houndslow barracks of the Seventh Hussars made headlines for two weeks in the summer of 1846. White had been sentenced by a court marshal to 150 lashes for insolent behavior while intoxicated, and he died some five weeks after the punishment had been inflicted in its full measure. The regimental surgeon, Dr. James Warren, certified White's death as due to an inflammation of the heart, but on receiving information about the circumstances of White's death, Wakley, in his capacity as local coroner, determined to hold an inquest. After an initial round of testimony on July 20, which included the opinion of Dr. Warren and his colleagues that White had died of natural causes, the inquest adjourned so the noted surgeon Erasmus Wilson might make an independent medical examination.[78] When the inquest resumed a week later, Wilson promptly contradicted the earlier medical testimony. He identified White's fatal disease as inflammation of the heart and lungs caused not by any of the theories put forward by the medical men connected with the military's initial inquiry (variously attributed to a change of temperature and exposure to cold, confinement to the hospital, and the depression of spirits consequent on flogging), but instead by a pronounced "muscular disorganization" around the spinal area brought on by the flogging.[79]

Wilson explained his disagreement with the previous medical testimony by reference to his own specialist credentials, his testimony founded upon "a scientific observation connected with pathology, and one which has never been made before."[80] In his summation Wakley supported Wilson's claim to distinction, instructing the jury as to his eminence in the field and his reputation for "unimpeachable veracity" and only then observing that the jury "had to choose their own authority, to select between Mr. Wilson and the three other, also eminent medical gentlemen." The jury apparently chose Wilson; it found that White had died from the effects of a "severe and cruel flogging" that had nonetheless been ordered by a "duly constituted" military tribunal. To this verdict of formal legal exoneration, however, the jury appended a rider "expressing their horror and disgust" at the continuing legality of this "revolting punishment" and voicing their hope that White's death would lead to a petition campaign "for the abolition of every law and regulation which permits the degrading practice of flogging to remain one mo-

The Flogging Excitement
AT HOUNSLOW.

From London resounded to Newry,
The fate of John White, the Hussar.

The Middlesex famed gallant jury,
In history recorded shall be,
They struggled together like fury,
For the good of the army we see,
Three times a strict investigation,
To Heston they went from afar,
To come to a determination,
Respecting poor White, the Hussar.

John White was a native of Yorkshire,
Brought up in the famed town of Leeds
He had been a policeman and soldier,
Though scarce in his prime as we read
He died by the laws of his country,
With his body all covered with scars,
May never again a brave soldier,
Die the death of John White, the Hussar

You sons of Great Britain attention
Pray give unto me for awhile,
Let us hope that on every brave soldier,
Dame Fortune in future will smile ;
This disgraceful affair now at Hounslow,
Has great excitement caused afar,
The death of John White, the brave soldier
Of Her Majesty's 7th Hussars.

May this Hounslow affair be a warning
To all Generals and Colonels afar,
And think of the fate night and morning,
Of John White, the poor gallant Hussar

John White was tied up to a ladder,
No halbert there was in the place,
It caused his comrades to shudder,
To England it was a disgrace ;
The cat on his shoulders did rattle,
The sound it did echo afar,
Oh ! remember the Hounslow battle,
And John White of the 7th Hussars.

Round Isleworth, Brentford, & Hounslow,
And Heston, it caused much pain,
In Twickenham, Richmond, & Hampton,
In Sunbury, Egham, and Staines ;
Thomas Wakley empanelled a jury,
Which caused great excitement afar,

There is no man more brave or bolder]
Or in battle more glorious will stand,
Than a stout-hearted true British soldier
Who fights in defence of his land ;
He struggles for honour and glory,
And falls with a smile in the wars,
Then why should we tell such a story,
Of White, the poor 7th Hussar.

Then Britons all meet in communion,
Petition the State and the Queen,
Be ready, be willing, and soon then
To banish such disgraceful scenes ;
May flogging be ever abolished,
At home and in nation's afar,
No more let a soldier be punished,
Like White, of the 7th Hussars.

Tied up hands and feet to a ladder,
While the sound of the cat reached afar
Oh, Britain thy deeds make me shudder,
Remember poor White, the Hussar.

BIRT, Printer, 39, Great St. Andrew Street,
Seven Dials, London.

Ballad composed in celebration of the Houndslow inquest into the death of Corporal White, 1846. Courtesy, British Library.

ment [*sic*] a slur on the humanity and fair name of the people of this country."[81]

Reaction to the verdict and rider was generally favorable. Wakley and his jury were feted in the pages of the popular press, in street balladry, and at public meetings such as the one convened at the

National Hall, Holborn, which concluded with a resolution "that
the searching manner in which that investigation was conducted reflects the greatest credit on the coroner, Thomas Wakley, Esq., MP,
and the public-spirited verdict of the jury, condemnatory of the cruel
torture, and demanding that such atrocities should be put to an end,
is deserving of all praise and public support from the people of
the United Kingdom."[82] Several papers concentrated their remarks
on the wider implications of the case, drawing particular attention
to the propriety and significance of the jury's rider. The *Morning
Chronicle* praised the jurors' expressed wish that the case would
spark a national movement for the abolition of corporal punishment, seeing nothing unusual in having inquests serve to galvanize
public opinion in this manner. The "Houndslow Inquest" augured
an end to military flogging, the *Chronicle* ventured, since "public
opinion—never so strong or so conscious of its strength as at the
close of the present session of 1846, and never, we may add, so sensitive to a social abuse and a legalised immorality as at the advent
of the present Ministry of social reform—will no longer tolerate that
hateful regime of the lash."[83]

But it was *The Times* that made the most considered effort to align
the popular inquest with the normal course of reform, in effect to
confirm the inquest as conceptualized by Wakley and others as a
matter of political principle. This came in response to a letter to the
editor arguing that any legislation resulting from the case ought not
to be based on "the fallacious test of a solitary accident which due
care would have prevented, still less by that idle clamour miscalled
public opinion, but by the arguments of competent judges, medical
as well as military." *The Times* rejected this objection out of hand.

There is nothing unnatural or extraordinary in the circumstance that a
particular incident should forcibly concentrate public attention on a system, and precipitate a decision on its merits, which might otherwise have
remained, not indeed unsuspected, but unquestioned in any serious tone
for some time to come. Such is the well-known history of most of our
great reforms. Some striking example of vicious principles or defective
legislation has struck home to the convictions of the people, and aroused
them to exertions of thought and action which would hardly have been

called forth by abstract argumentation, but which insured a patient hearing and deliberate consideration for such plans and petitions as would otherwise have been overlooked or overruled.[84]

There was a clear and legitimate relationship between sensation, exposure, and reform. Sensational cases brought sensation to an end. There was nothing inherently radical in the inquiries of the popular inquest. Rather, it constituted a part of a progressive, self-improving social body.

The Times's domestication of the inquest, subsuming it under the normal process of reform, also indicated its limits as a tool of full-fledged radical politics. It was vulnerable, first, to criticism from the most resolute enemies of "Old Corruption"—those involved with the unstamped and Chartist press, to take the most obvious documentary example—who, across the full spectrum of issues, expressed a profound alienation from what they considered the increasingly accommodationist line taken toward the post-1832 political establishment by their erstwhile allies among the ranks of the middle-class professions.[85] Under this kind of critical lens, the inquest appeared either as one in a series of victims to the treacheries of the new order or as an irrelevance, mired in a kind of humanitarian fuzziness that obscured the fundamental issue of rights and wrongs. An example of the former diagnosis comes from the leader columns of the *Champion and Weekly Herald,* where an account of an 1838 Yorkshire workhouse death opened with a familiar, post-lapsarian vision of the inquest: "We have more than once called the attention of our readers to the flagrant neglect into which these courts have fallen: one of the most ancient and important offices of the land is suffered to become a mere mockery; a blanket to cover up and smother justice, excepting always in such cases as do not furnish any thing reproachful to men in power."[86]

Nor was this merely an attack on a pre-Wakley enclave of coronatorial corruption. Wakley himself could be subjected to the criticism, both for his general political actions (the *Champion* professed itself "by no means satisfied with the conduct of Mr. Wakley respecting the Poor-Law") and for his handling of specific inquests.[87] In its response to the inquest verdict in the case of the pauper James Lisney, the Chartist *Northern Star* provided its readers with an example of Wakley's shortcomings as a coroner which doubled as an indicator of the inquest's general tendency toward the vague and the inconsequential. The jury in this case had attached, with Wakley's express support, a rider to the verdict stating "their opinion that it was not humane to imprison him without fire, and on low diet." "Humane!," the *Northern Star* retorted, "was it *legal,* was it constitutional?" The inquest's sentiment might soothe individual and collective consciences, but, by substituting sentiment for justice, it had lost its political efficacy. "The Coroner's inquest, by the latter part of this verdict," the *Star* lamented, "left in doubt of what offence the Guardians were guilty."[88]

From the other side, the inquest's radical edge was dulled by familiarity. By the middle of the nineteenth century, proclaiming the inquest as a popular institution had become standard fare. Though cast as "eminently the magistrate of the poor," in the words of the (by then) centrist *Times,* the emphasis of such descriptions had noticeably shifted from reform based on exposure "from the root" to that based on "prevention"—prevention of a relapse into a former state of abuse, on the one hand, and of undue suspicion of abuse by the public, on the other. The real value of the inquest, declared Coroners' Society President Samuel F. Langham in 1865, lay in its capacity to secure the fruits of a progressive liberal polity, a claim that he illustrated by recourse to the tried-and-true example of prison reform. As a result of the efforts of an earlier generation of reformers spearheaded by Howard, Langham observed, the modern prison regime was benign (i.e., beyond suspicion). But he insisted that this did not mean that inquests into prison death no longer had an important function. "The coroner is now called upon to be the watchful guardian of the public," he explained, "to prevent a relapse into the oppression of the past."[89]

The lineaments of Langham's watchful inquest were further articulated in papers delivered the following year at the National Association for the Promotion of Social Science (NAPSS), urging that workhouse inquests be statutorily required. The workhouse had not yet been subjected to the corrective effects of humanitarian reform, the barrister Henry Cartwright observed, and thus now "specially demands the guardianship of public opinion" secured by that "valuable tribunal for the protection of the weaker members of the body politic from the oppression of power and wrong-doing"—the coroner's inquest.[90] The Liverpool medical officer of health, Joseph Pope, ventured that, if workhouse inquests were held in all cases, they might soon become necessary, as in the case of contemporary prison inquests, solely as protective measures. Inquests only "exposed" in the early stages of a reformist trajectory, Pope observed, explaining that "the cause of calamities being discovered, it then falls to the lot of others to remedy it, and to avoid repetition."[91] A final, and most obviously "conservative," function of the modern inquest was broached in comments made by A. O. Charles. In the typical prison inquest of their day, Charles informed his NAPSS colleagues, "the Coroner usually charges the jury to the effect that the inquiry is called for, not because of any suspicion of neglect or unfair play, but for the purpose of conveying to the public an assurance that though a prisoner, and shut out from the attention and sympathy of his own family and friends, every attention and care had been bestowed on the deceased."[92] It was the legitimate and socially necessary duty of the modern inquest, in other words, to reveal to the public not the existence of abuse, but its absence. The inquest properly addressed vestigial concerns about "Old Corruption" in an era in which corruption had been effectively banished from the scene.

As a matter of principle, exposure as a legitimate aid to reform and to the maintenance of public confidence in institutions and practices falling under the regulatory penumbra cast by the modern state was confirmed legislatively in the second half of the century. From the 1860s onward, parliamentary provisions for regulating institutional spaces like inebriate homes, infant-minding establishments, lunatic asylums, and certain workplace environments named the inquest as a supplementary investigative mechanism in cases of fatality.[93] In an age of equipoise, the inquest seemed to have found its niche.

Still, a tension remained between the inquest as a tool of exposure and demonstration, on the one hand, and of procedural inquiry and explanation, on the other. This is discernible in the conflicted view of the inquest taken by later radicals like G. W. M. Reynolds and Charles Reade, who at once embraced the principle of the popular inquest and criticized its tendency toward conventionality.[94] Reynolds's newspapers, of course, fed off a steady diet of tales of individual suffering, deprivation, and injustice brought to light by inquests and were eager to transmit and elaborate upon the humanitarian narratives engendered by such proceedings. Under the banner headline "Revolting Workhouse Revelations," for instance, *Reynolds's Newspaper* of 14 November 1869 treated its readers to an instructive story of workhouse neglect drawn from the Saint Pancras coroner's court, observing that "the details evoked . . . make one shudder to dwell on" and predicting that "our numerous readers, we are sure, will cry 'Shame!' when we have recorded the shocking story." But Reynolds's papers also expressed exasperation and even outrage at the limitations of the contemporary inquest. Coroners' juries, one regular columnist observed, were ineffectual placeholders, their opinions dismissed as "the mere good-natured expressions of lazy-minded men, who will not look at the root of the baleful fruit which they abhor."[95]

For his part, the celebrated sensationalist author Charles Reade often cast the inquest in a resolutely heroic light in his fictional exposés. His novel depicting abuse in the national prison system, *It Is Never Too Late to Mend* (1856), favorably contrasted the useless "gullibility" of official inspection to that offered by the searching scrutiny of the popular inquest: "When twelve honest Englishmen, men of plain sense, not men of system, men taken from the public not from public offices, sat in a circle with the corpse of a countryman at their knees, fiebat lux; 'twas as though twelve suns had burst into a dust-hole. 'Manslaughter!' cried they, and they sent their spokesman to the mayor and said yet more light must be let into this dust-hole, and the mayor said 'Ay, and it shall too. I will write to London and demand more light.'"[96] Yet Reade's relationship to the inquest as a tool for exposing institutional abuse was more problematic in practice. In his role as investigative journalist for the *Daily Telegraph* and the *Pall Mall Gazette,* he delivered excoriating judgments on specific inquests that fell under his notice. His comments on the remarkable and highly publicized spate of deaths from broken ribs among inmates of public lunatic asylums in the 1860s and 1870s evinced a clear and at times hostile skepticism toward the value of inquest proceedings.[97] Such deaths, Reade wrote in an extended letter to the *Pall Mall Gazette,* were matters of grave public concern, but public knowledge of them was wholly inadequate, dependent upon "the faint light of an occasional inquest, conducted by credulity in a very atmosphere of mendacity." Reade proceeded to detail the results of his own investigations into several recent "pulverization" cases at asylums, results often directly critical of inquest findings. The verdict of a Middlesex jury inquiring into the death of Matthew Geoghegan at the Hanwell Asylum in the early 1860s, for instance, provoked a vintage Readean tirade: "With their eyes to confirm what their ears had heard warn, and their ears to confirm what their eyes saw written on the mangled corpse," the jury had found no evidence to show the cause of death, "as direct a lie," he foamed, "as ever twelve honest men delivered."[98]

The light emanating from inquests at midcentury thus produced contrasting effects as it was refracted through a range of differing interpretive prisms. Plucked from relative obscurity by its imbrication with the early nineteenth-century political discourse of popular liberties, the inquest's "public" mandate had been secured in principle but was ambiguous in its significance, alternately radical and conventional, heroic and passive. In this protean form the inquest came up against a second powerful and avowedly modern framework for its actions, one that in many respects was coextensive with the inquest's historically derived political rationale and yet in others was based on a different, ostensibly apolitical, foundation. It is to the triangulated relationship between the "traditional" inquest, popular inquiry, and the investigative ideals promoted by scientific medicine that the discussion now turns.

REGISTERS OF DEATH

Inquests and the Regime of Vital Statistics

At the same time that the inquest was being recast as a traditional bulwark of popular liberties, it was being infused with the self-consciously forward-looking ideology of science in the service of a modern social order. A new compound vision of the inquest—as a medically driven tribunal guarding the interest of the people—emerged out of an intense reformist activity during the first half of the nineteenth century, activity that constituted a singular moment of convergence in radical medicine and politics.[1] Though the details of this coexistence would be revised at different moments over the course of the century, the underlying assumption of a link between a progressive medicine and the inquest's political mandate as a popular tribunal of openness retained its force.

This chapter considers this process in the context of two distinct but related institutional mechanisms for managing death, inquests and the civil registration of death. Both were exercises in interpretation and translation, in making meaningful statements about death. Yet they operated on different registers: the inquest focused on the individual case, whereas death certification focused on the derivation of aggregate figures unencumbered by individuating circumstances. Each produced its own characteristic narrative, humanitarian in the former case and statistical in the latter. It was clearly as a unit of statistical knowledge that death spoke most directly to the needs and sensibilities of a modern legislative agenda increasingly focused on an abstracted and biological conception of population. It was less certain whether the office of the coroner—even one modernized and medicalized—still had a useful role to play in making death publicly meaningful.

I

When he became Middlesex coroner in 1839, Wakley the Radical stormed the citadels of "Old Corruption." At the same time, Wakley the medical reformer waged a related, but less obviously heroic, campaign to redefine the recognized criteria for submitting categories of death to official—and, ideally, medical—scrutiny. Ultimately, his objective was to have all cases of unexplained death referred for consideration to a local, medically trained coroner. Having used the editorial pages of the *Lancet* for over two decades to promote the cause of a medical coronership, Wakley quickly took advantage of the discretionary powers vested in him as the newly elected coroner for West Middlesex to influence practical procedures for notification. He began this latter effort in typically controversial fashion when, in his September 1839 instructions to parish officers, he sought to secure a broad remit for inquest investigations by insisting that he be notified of "all cases when persons die suddenly."

By claiming that the ancient category of "sudden death" fell within the coroner's investigative competence, Wakley was swimming against a strong judicial current. According to the earliest authoritative statement on inquest practice, the thirteenth-century *Officium coronatoris,* it was one of the coroner's primary duties to "go to the place where any be slain, or suddenly dead, or wounded" and make inquiries.[2] Yet by the early nineteenth century a series of high court decisions had developed a restrictive interpretation of the coroner's responsibilities in cases of the "suddenly dead." *R. v. Kent JJ.* (1809), the most important of these rulings, centered on the attempt by the Kent County magistrates to disallow inquest fees in a case where there had been "no ground to suppose that the deceased had died any other than a natural, though sudden death."[3] During the high court's review of the Kent case, conservative Chief Justice Ellenborough endorsed the magistrates' use of their powers of financial scrutiny as a way of ascertaining whether (and ensuring that) inquests were "duly taken." This constituted a salutary check, in Ellenborough's estimation, against the "many instances of Coroners having exercised their office in the most vexatious and oppressive manner, by obtruding themselves into private families to their great annoyance and discomfort, without any pretence of the deceased having died otherwise than a natural death; which was highly illegal."⁴

Though not amounting to a firm declaration of law, Ellenborough's support for what he depicted as the magistrates' public-spirited stand against an overly "inquisitorial" inquest was frequently cited, not only in subsequent judicial decisions, but also in authoritative works on inquest law like Sir John Jervis's 1829 *Practical Treatise*.⁵ By the time of Wakley's election, the strictures against inquests into sudden deaths without suspicion of violence or unnatural causes had become commonplace, with a growing number of county benches considering themselves perfectly competent to disallow inquest fees on these grounds alone.⁶ As defined by the magistrates sitting at quarter sessions and enforced by their hold on the purse strings, the inquest was allocated a highly circumscribed, supplemental role within the expanding system by which criminal violations of public order were policed.

Wakley's election as medical coroner and his instructions regarding sudden death alarmed supporters of the magistrates. To them, Wakley seemed to augur a new form of inquisitorial excess, one driven not by crude avarice—the charge commonly laid at the door of other "overreaching" coroners—but by the more dangerous aggrandizing calculus of professional zealotry. At risk from Wakley's uncompromising pursuit of medical inquiry, they suggested, was the integrity of both the public purse and the public body.

By casting his appropriation of sudden death as an exercise in medical self-interestedness, Wakley's opponents were tapping into a culturally profound and socially diffuse concern at medicine's perceived transgressions against the physical integrity of death. As Ruth Richardson has forcefully shown, dissection and the means used to procure a supply of subjects for this purpose were at the core of a mounting anxiety over the proper relationship between the living and the dead in the first half of the nineteenth century.⁷ Unlike the contemporary French case, where brimming hospitals seem to have provided surgeons with an ample stock of corpses from among the ranks of the sick poor, the rapidly expanding British market for hands-on, anatomically based medical training in the opening decades of the nineteenth century was facing an acute shortage of bodies. Historians have explained this in part by the fact that, in both popular and polite understanding, dissection had long been seen as primarily a ritual of violence visited in accordance with the law. From the reign of Henry VIII until the passage of the Anatomy Act of 1832, the only sanctioned supply of corpses for medical dissection had been a fixed annual number of hanged felons.[8] In the early part of the nineteenth century, political and medical reformers waged a sustained campaign to secure a more regular supply of bodies, a cause invested with mounting urgency by reports of the extraordinary means taken to bridge this market gap, including the flourishing trade in body snatching practiced by the notorious "resurrection men." The end result was the 1832 Anatomy Act, which made the unclaimed bodies of those dying in pauper institutions available to anatomy schools. But, as Richardson observes, passage of the act by no means diffused the dissection controversy and, alongside protests against other items in the brace of utilitarian-inspired legislation passed during the immediate post-Reform era, it continued to serve as a lightening rod for social tension.[9]

When Wakley issued his 1839 directive respecting sudden death, a largely hostile daily press immediately tapped into this rich vein of cultural anxiety. Under banner headlines like "Wholesale Post-Mortem Examinations," editorials took the order as evidence of the new Middlesex coroner's unsavory *"predilection* for dissecting" and his seeming willingness to flaunt public standards of decency by *"mangling without reasonable motive."*[10] Several explanations for this outrage against the public accompanied the reports. In the *Observer*'s judgment, the Middlesex citizenry was reaping the bitter harvest of electing a medical coroner more concerned with pressing his own narrow professional agenda than with either the public good or the good of medicine as a whole. "Mr. Wakley ought to know," it declared in its September 29 editorial, "that an inquest can neither be held for mere pleasure nor his mere profit, nor to enable him to sit in judgment on his brother medicals who are in practice, nor to show off his own surgical learning." The *Morning Advertiser* denounced the scheme as another instance of the familiar contemporary tendency toward a distorted economics of exchange, through which waste was converted into unwholesome profit. "As the bones of horses and cattle are now converted into manure," the *Advertiser* complained, "*so,* IN FUTURE, *are the last dear remnants of the dead to be converted into a mode of* INCREASING THE FEES OF CORO- NERS!"[11] The *Morning Herald* added cynical electoral politics to the explanatory mix. Wakley's justification of his activist and interven- tionist stance at a recent inquest (claiming, according to the *Herald*'s report, that "the office of the coroner rose from the people—it was instituted for the protection of the people—the coroner himself was appointed by the people, and it was his duty to watch over and pro- tect the interests of the people") was a naked attempt at "combin- ing his electioneering with his official labours—converting, in fact, his Coroner's court into a district committee pro tem . . . This vil- lainous iteration about 'the people,' 'the people,' 'the people,'" the *Herald* concluded, "has very little to do with the 'Crowner's quest,' but it has a vast deal to do with the next election for Finsbury."[12]

Wakley's most powerful opponents, of course, were the Middle- sex magistrates themselves, with whom he was locked in perpetual combat until his death in 1862.[13] The Middlesex magistrates ini- tially responded to Wakley's parish directives with outrage and re- solved to convene a committee of inquiry into his activities. When the committee's report was delivered at a meeting of the quarter ses- sions in October 1839, however, it was muted in its direct criticism of Wakley, as the investigating committee had discovered, to its con- siderable surprise, that both his inquest numbers and his rate of postmortems were in fact lower than those of either his legal prede- cessor (Thomas Stirling) or his legal colleague (Baker).

The committee's findings, belying as they did the images of untrammeled postmortems circulating in the newspapers of the day, reflected among other things Wakley's own sense of having to proceed cautiously in treacherous waters. As a series of *Lancet* edi- torials in the early years of his tenure as coroner made clear, Wak- ley believed that his handling of inquests would be taken by the public at large as a test case for the principle of a medical coroner- ship.[14] Not that he or any other (informed) observer of inquest prac- tice could doubt that "the power and utility of the inquest are greatly diminished in consequence of the rarity of post-mortem examina- tions."[15] It was, he asserted in the *Lancet,* a virtual medical truism that "in the vast majority of sudden deaths the cause of the calamity is altogether hidden, entirely obscured from view."[16] "Were we to be asked whether we believe that the utility of the Coroner's inquest would be increased by the institution of the postmortem examination in all cases of sudden death," another *Lancet* editorial declared, "we should emphatically reply *Yes.*"[17]

But inquests, far from offering a clear ground for instituting medically maximizing reform, operated in a social and historical context in which medical and public sensibilities were decidedly out of sync. At the most prosaic level—though hardly unimportant in Wakley's estimation—was the question of cost: "JOHN BULL is an ardent lover of justice," he mused, "but he talks of it with his hands in his breeches-pockets."[18] Suspicion of financial violation, however, paled in comparison to its physical analogue. Having only recently freed dissection from its association with scaffold rites and now battling with its post–Anatomy Act image as a ruthless and calculating violator of the poor, the medical profession had to be alert to both the visceral fear of dissection and the associated cultural stigma attached to it.[19] In Wakley's estimation, these worked in a paradoxical relationship: so long as a stigma attached to autopsied bodies, a comprehensive system of inquest postmortems would generate public distress; yet the only way to remove the stigma was to make postmortems a routine feature of the coroner's inquiry.[20] This was a knot that only time and incremental experience could begin to disentangle. The public had to be won over to the cause of searching medical inquiry by force of example, and it was his own court that Wakley and others hoped, over time, would accomplish this.

Recourse to a couple of Wakley's early cases illustrates his pedagogic method and points up the limits of his intended lesson. The first indicates a degree of success enjoyed by Wakley in exploiting his radical credentials to promote the cause of medical inquiry. Mary Lewis, a fifty-seven-year-old widow, died unexpectedly in early October 1839. Her death "presented no feature of interest or importance, as far as the evidence went," in the view of the *Morning Chronicle* reporter, but her attitude both to her own impending death and to the broader debates over the inquest's public duties made hers a death worth reporting. According to the deposition given by her son, Lewis's chronic complaint had baffled her medical attendants, and this had led her to regard her body as the keeper of a medical mystery. "She repeatedly expressed a wish that at her death her body might be opened," her son informed the court, going on to suggest that his mother's belief in the benefits of postmortem investigation was in part a response to Wakley's battles in the name of the popular inquest. She had been following the controversial September inquest on the Hendon pauper Thomas Austin (see ch. 1) and on the very night of her death had proclaimed Wakley "an excellent man." Though declining to act on Lewis's exemplary attitude in what was a clear case of natural death, Wakley took the opportunity to express his desires for the future: "The case was one in which a *post-mortem* examination of the body would have been useful, and he hoped the time would come when they would be made in all cases of sudden death."[21]

The case of Julius Thomas Pampe, inquired into by Wakley's court in July 1839, on the other hand, suggested that the twin ideals of medical knowledge and public instruction were separable propositions. Pampe had drunk a pint of beer with his Sunday dinner, gone to bed (according to his daughter) in good health, and was dead by morning. As no unusual circumstances had been elicited from the "respectable" family and the other lay witnesses, the *Morning Herald* report proceeded directly from the assurances of the deceased's daughter that her father had "no doubt died a natural death" to Wakley's summation. "There can be scarcely any doubt as to the cause of death," Wakley instructed the jury. But, he continued, "if you have any, it can be removed by causing a post-mortem examination of the body to take place." Though not strictly necessary in the interest of criminal justice, Wakley admitted, there were other benefits to be attained by a publicly performed expert inquiry into such a mysterious death. Postmortems in all cases of sudden death were beneficial, not only because they would result in a medically sound verdict "without a shadow of a doubt," but also, Wakley advised his jury, because "if any of you were afflicted with the same symptoms under which the deceased laboured you would, if it were discovered that the supper and porter he took accelerated his death, have an example whereby to regulate your own diet."[22]

A medically centered inquest held out the promise of a fruitful encounter with the public as both subject and object of its investigative operations. Though the ostensible reason for the jurors' presence was to decide on the cause of a fellow citizen's untimely death, they were also there to learn, to project themselves into the story being unfolded, to regulate themselves according to the errors therein revealed. The corpse was ideally an object lesson, decoded and delivered to the public who, in turn, were to subject themselves to the knowledge inherent in the lesson. According to this model of death inquiry, inquests were valuable not merely (or indeed even mainly) at the level of literal policing. Instead, the inquest's proper remit was at the same time to enact a searching investigation in public to deter would-be criminals, to translate death into a public lesson in mortality, and to defuse anxieties that a death was something other than natural. With this in mind, Wakley urged the public and coroners alike to sanction a searching inquiry "in *all* cases of strictly sudden death, and to repudiate with dismay and indignation, the pernicious and disgusting doctrine that they are not to call the functions of their office into operation unless, in the first case, there be a *well-grounded suspicion* of wrong-doing."[23]

But the Pampe case also illustrates the limits to the accommodation between the popular and the medicalized inquest. After a lengthy disquisition about the ultimate need to achieve universal postmortems in such cases to avoid arousing selective suspicion,[24] Wakley placed the question in the jurors' hands. Their response was briefly yet tellingly reported by the *Herald* correspondent: "The jury warmly commended the views of the coroner, but having no doubt on their minds as to the cause that led to the present inquiry, they returned a verdict of 'Natural Death.'" While medical investigation was framed within a discourse of public interest, the level of the public's interest could not guarantee that the needs of medicine with respect to death would be served.

The available evidence on the rate of postmortem examinations over the course of the century bears out Wakley's experience of limited success in his efforts to promote a more comprehensive regime of medical scrutiny in such cases. Though it is possible to trace a gradual increase over the course of the nineteenth century, Wakley's aim of routine postmortems had to wait for the better part of a century.[25] In his court, to his chagrin, "natural death" might be a fully satisfactory answer to the questions posed within the terms of public inquiry.

II

Attempts to forge an alliance between popular and medical constructions of the modern inquest were a concern not merely of a leading medical coroner like Wakley, but also of key figures in a new, bureaucratically centered approach to public mortality, of whom Dr. William Farr, statistical superintendent at the General Register Office (GRO) and a formidable public health campaigner, is the best example. An enthusiastic backer of Wakley's 1839 campaign for the Middlesex coronership, Farr embraced Wakley's vision of an inquest set free from a strict relationship to crime detection and used his position at the GRO to promote this medical turn. "The Coroner's inquest is entirely a popular institution," Farr declared in his most sustained discussion on the contemporary state of medicolegal inquiry, one that both "engages the body of the people in the administration of justice" and assists them in preserving health by instructing them on "the action of general causes, such as nuisances, in destroying life."[26] Farr also supported Wakley in casting medically unexplained sudden death as a paradigmatic case for the modern inquest: in deciding on whether such a death was worthy of inquiry, Farr asserted, "the fact that the deceased has or has not been recently visited by a medical practitioner should be taken into account." This would require a significant number of inquests in cases where medical scrutiny in the end revealed a thoroughly ordinary cause—ordinary, that is, from the circumscribed model of policing adopted by magistrates and their supporters. Yet, like Wakley, Farr insisted that these were not wasted inquiries, if for no other reason than because they neutralized suspicion. Through such inquests, he observed, "the supposition of violence [is] negativated, and this decisive result is ample compensation to society for the expense."[27]

According to Farr, the social peace of mind resulting from attention to the local and highly individual nature of any given death was a legitimate—indeed, a crucial—aim of the inquest. "Its most important function," he stated in an unambiguous embrace of the particularistic vision of the inquest, "is to dissipate unfounded suspicions."[28] But Farr's support of Wakley in principle belied profound fissures in the new English way of managing public death made evident in the wake of the Births, Deaths, and Marriages Registration Act of 1836 (6 & 7 Will. 4, c. 86). For Farr, civil registration required inquests that produced not only medically credible verdicts, but also ones that were usable for the purposes of vital statistics.

The Births, Deaths, and Marriages Registration Act laid out the procedures whereby, for the first time in the nation's history, every death could be recorded by a central bureau of state, the GRO.[29] The task of organizing and interpreting mortality returns fell to the GRO's Statistical Department, which Farr ran between 1837 and 1880. Farr's objective for the emerging system was to ensure its capacity to produce a nationwide system of comparable mortality statistics, one that correlated the isolated observations of individual medical practitioners, thereby turning them into functioning elements of a coherent signifying system. Giving deaths their proper name lay at the core of this project: "The nomenclature," Farr asserted in the first of his annually published letters to the registrar general, "is of as much importance in this department of inquiry as weights and measures in the physical sciences."[30] Ideally, he continued, the nomenclature would guarantee and reflect a consistent and precise system of certification, one in which causes of death were "uniformly registered under the same names, and each cause of death designated by one word."

For Farr and his colleagues, civil registration of death was inseparable from its accurate and reliable medical certification. But the 1836 act was singularly ill-equipped to serve as a filter for collecting such units of pure knowledge. Provisions for medical certification were given low priority; indeed, as both John Eyler and M. J. Cullen have noted, the very idea of registering the cause of death (as opposed to its mere fact) was an addition hastily tacked on to the parliamentary bill.[31] From a statistical perspective, this resulted in flaws at every level of the derivation, registration, transmission, and recodification of causes of death. Causes of death, for a start, were not defined medically. In a case of death unattended by a medical practitioner, registration could proceed, subject to the local registrar's discretion, on information supplied by lay informants.[32] This placed the registrar—whose sole statutory qualification for office was that he was not a publican, an undertaker, or a debtor—in the position of deciding what constituted a sufficiently explained death on the testimony of nonmedical informants. Moreover, even when practitioners were involved in the registration procedures, their obligations remained vague. For instance, though the registrar was entitled to request information from a medical attendant in order to register a death, neither he nor his superiors at the GRO's headquarters in Somerset House could dictate the form in which such information was given.[33]

Then there was the problem of the inquest. In cases of death inquired into by an inquest, it was the final verdict that served as the registerable cause of death.[34] This meant that some 5 to 7 percent of deaths each year were certified according to the judgment of a public jury. In addition to its role as a source for vital statistics, the inquest was meant to serve as a supplementary resource in difficult cases. When a local registrar believed that the normal channels of information were unsatisfactory, he could bring a death to the attention of the coroner. It was then the coroner's responsibility to decide whether the death merited an inquest.[35] But, from the perspective of statistical purists, this merely engulfed the registration system in an epistemologically suspect world in which parish beadles, acting as coroners' officers, conducted preliminary inquiries, and lay coroners, devoid of any medical information, decided whether or not the case merited an inquest. Nor did it really matter which course the coroner chose: if he decided against an inquest, the registrar would have to enter the cause of death to the best of his knowledge and register it as "uncertified"; if an inquest was held, the cause of death was registered according to the whims of a lay jury.

The Registration Act seemed to leave the certification of death as a discretionary and heterogeneous system capable of producing at best a promiscuous mixture of accurate medical causal explanation with unusable lay opinion. Though in statistical terms the contributions of inquests to the registrar general's annual returns were marginal, in discussions about the present and future shape of the mechanisms of civil death management, in particular, and of the objectives of state medicine and public health, in general, the inquest figured prominently as the emblem of the system's weaknesses. It was precisely the minutiae that threatened the whole system—returns, in order to signify, had to be pure. "The smallest evil of a system of registration, which admits of erroneous entries," a leading medical statistician warned, "is the vitiation, *pro tanto,* of all deductions from the death returns."[36]

For those already urging the necessity of a medically directed inquest, the solution was simple. A reformed inquest, they insisted—one placed in the hands of a medically trained coroner whose professional sensibilities predisposed him toward viewing any medically incomplete account of death as sufficient grounds for an inquest and whose inquests regularly involved searching investigation and the testimony of reliable medical witnesses—could serve as a reliable supplement to certification. Civil registration, in this sense, was co-extensive with the inquest. It was the inquest's founding assumption that "the State has a right to ask how every Englishman dies," the *Lancet* proclaimed in an 1839 editorial welcoming the GRO's first annual report. Civil registration was merely a new variant on this ancient principle, one attuned to the needs of modern governance but subsumed under the inquest's institutional example. "The registration of the causes of death may be regarded as a partial extension of the principle of the inquest to the entire population," the *Lancet* continued. "By stating the causes of death in ordinary cases, and conscientiously indicating cases of sudden death and suspected violence, the medical practitioner will contribute essentially to public security, the conservation of life, and to the progress of medico-statistical science."[37]

But the *Lancet*'s version of two complementary and thoroughly up-to-date mechanisms for managing death struck others as overly optimistic. Even such outspoken supporters of the inquest as William Farr found themselves confronted with a tension between the narrative objectives of inquests and vital statistics. As Farr observed in his writings on the methods of his science, statistical knowledge depended on eschewing the particular. This presented the vital statistician with an enabling limitation on his work. True, his investigative domain was "the gloomy kingdom of the dead, whither have gone in 20 years 900,000 English children, fathers, mothers, sisters, brothers, daughters, sons, . . . each having left memories not easily forgotten; and many having biographies full of complicated incidents." Yet, Farr observed, this had no bearing as far as mortality returns were concerned: "Fortunately for this inquiry, they appear divested of all colour, form, character, passions, and the infinite individualities of life: by abstraction they are reduced to mere units undergoing changes as purely physical as the setting of stars of astronomy or the decomposing atoms of chemistry."[38]

The complex play of identities, histories, and passions that inevitably followed a person—considered from his or her own life story—into death thus threatened to render that death meaningless from an analytical point of view. A national system of certification promised an ordering mechanism through which individual death could take on universal significance. In the registrar general's analytical tables, a death lost its limiting individuated form and became a unit of comparative data, which both supported and gained new meaning from a structure of statistical meaning. With evident relief, Farr held that, in the name of useful knowledge, it was the task of all participants in the death management and certification apparatus to extract from the chaos of the individuated and the particular the kernel of comparative significance.[39]

However, as Mary Poovey has observed, this ideal of abstraction in statistical narrative was by no means settled in the early days of the civil registration system. Indeed, her analysis of debates over statistical method at the British Association for the Advancement of Science in the 1830s reveals a tension between analytical ("figures of arithmetic") and individuated ("figures of speech") approaches to statistics operating both between statistical theorists and within statistical texts themselves.[40] These figurative levels operated side by side in contemporary medical texts, none with more dramtic effect than Augustus B. Granville's 1854 *Sudden Death*. Granville's text combines a stated intention to utilize the GRO's annual returns for a scientific analysis of sudden death with passages that, by allowing simple factual units inscribed in the analytical tables to transmute into an opaque and pathetic materiality, repeatedly transgresses

Farr's strictures against the chaos of individuation. After praising the tables' medical descriptions of death, Granville proceeds at some length imaginatively to reconstruct a series of tragic tales to which the profiles referred. He is overwhelmed and abruptly halts his ostensible digression with the following confession:

> But enough! the heart sickens at these dreadful spectacles of human infirmities of body, immorality, insanity, and crime. Often as I went on scanning these stern and truthful details, the scenes themselves so forcibly recorded, were brought before and appeared to confront me,—until the very air of the subterranean chambers in which I was writing, in the doubtful twilight of an expiring day, seemed reeking with cadaverous effluvia, and the rattling of bones was heard along the dark vaulted passages.[41]

Farr's emphasis on abstract codings, even if at times difficult to put into practice, highlighted a real and obvious tension with the inquest's narrative agenda. Inquests took an individual case with unnatural or suspicious overtones and sought to determine its proper alignment within the network of factors that marked it off as a special case. In deaths attributable to accident or violence, the disjuncture was minimized, since the registrar general's statistical tables provided columns under precisely those headings. The problem, instead, was with deaths classed as natural. A verdict of natural death answered two important questions posed by an inquest: it declared the suspicion that had instigated the proceedings unwarranted, and it allowed for a return to good public order. Certification in the service of vital statistics, on the other hand, required the inquest to satisfy a particular constituency of interested curiosity, to go beyond the socially useful explanatory threshold and answer the further question: In what sense—in what medical sense—was the death "natural?"

By the 1860s, the hope that the inquest could adapt itself to the expanding needs of state medicine had begun to yield to proposals favoring a more thoroughgoing reform. This development was in one sense illustrative of broader trends in the contemporary English public health movement. The 1860s and 1870s witnessed a decided shift in the movement's tactics and leadership, during which time medical men asserted their right of influence over the sanitarian agenda.[42] The rise to prominence of the research-oriented Dr. John Simon, first as Edwin Chadwick's replacement as head of the General Board of Health and later as the chief medical officer to the Privy Council, signaled, among other changes, a new insistence on statistical detail in the joint interest of the nation's health and of medical knowledge. From outside government, advocates of a more rigorous system of state medicine turned newly founded organizations like the National Association for the Promotion of Social Science (NAPSS) into effective policy-pressure groups for pushing the agenda of a nationally coordinated system of public health.

Henry W. Rumsey emerged as one of the most influential of these new-style sanitarians. Rumsey, a private practitioner from Cheltenham, is credited with introducing into the lexicon of English sanitarians the term *state medicine*. This term had a distinctively foreign pedigree at the time, redolent of the Prussian-styled centralized and bureaucratic health administration commonly set in opposition to the piecemeal and "archaic" English approach.[43] Rumsey's "Continental" sympathies, predictably, made him impatient with suggestions that the fortunes of death registration should be tied to inquest reform: "We have been told of the invaluable protection afforded the public by this ancient institution; but when all the froth of so empty a boast has subsided, the plain fact remains that a very large proportion of coroners' inquests leaves the causes of death wholly unexplained. Many a coroner seems still to be satisfied with returning the old slovenly verdicts, as, for example, 'Died from natural causes.'" Dispelling admiration for the inquest's peculiarly English lineage was a prelude to a more radical proposal for reform, as was made clear in his subsequent profession of doubt that "appeal to the coroner and his court, in all obscure cases throughout the country, would solve each embarrassing question as satisfactorily as prompt personal inquiry by a medico-legal officer."[44]

By invoking "a medico-legal officer" serving as a link and a filter between local registrars and national returns, Rumsey underscored his intention, and that of like-minded reformers, to create a new structure to shore up death certification procedures. His proposal would take the certification of death by registered medical practitioners as the basic units of statistical fact. All such certificates would be reviewed by the public certifier, however, enabling him to detect sloppiness or error on the part of local medical men. Furthermore, in cases involving no explanation of an authorized medical kind, the public certifier, and not the coroner, would be charged with investigating the case. If the death was unnatural or suspicious, the coroner would be called in to hold an inquest, but if upon inquiry the death was found to be natural, its medical cause would be registered as certified on the basis of the certifier's inquiry alone.

The proposal for public certifiers, echoed in various guises throughout the last quarter of the century, was deeply controversial. Ordinary medical practitioners bridled at the prospect of subjecting their death certificates to the inspection of an outside "expert" unfamiliar with the case histories, while coroners regarded with suspicion any proposal that encroached on their jurisdiction over inquirable bodies.[45] Rumsey and his supporters dismissed these objections, however, by arguing exigencies of modern governance:

> Year by year a most careful and elaborate abstract of the alleged causes of mortality, under the direction of the most scientific statist of the age, is presented to the public . . . The products of these remarkable analyses are employed in many special investigations into the causation and distribution of disease, in many callings and conditions of life, and in each district of the kingdom. From these again originate a variety of administrative suggestions, leading sometimes to new measures of medical police, legal inquiry, and public hygiene. How immensely important that the original units—the elementary atoms of the complex structure—should be real and true, that the 'facts' should be facts indeed! Otherwise the vast building rests on a foundation of sand, and its ruin is inevitable.[46]

It would be decades before this "immensely important" matter received the legislative attention anticipated by Rumsey.[47] The passage of time, however, did nothing to dampen concern. The 1893 Select Committee on Death Certification was a monument to the anxieties built up in previous decades about the laxity of the certification system and its implications for public order, public safety, and medical knowledge. In its fourteen days of deliberations, the committee heard virtually every aspect of English death management called into question. The select committee was treated to tales of secret homicide masquerading as natural death; of bodies buried or, worse, cremated without any ascribed cause; of causal explanations solemnly entered in the national ledgers of deaths when those explanations had no connection to any actual cases of death; of the systematic use of fictitious causes of death for fraudulent ends; and of the resurgence of the longstanding specter of premature burial, now causally tied to the inadequate provisions for medically verified death.[48] The impression left by the testimony was of an England awash with people who had been mislabeled in life or death, who embodied the dangerous gap between a seemingly reliable story and its actual referent.

Inquest verdicts that did not accord with the twin requirements of civil registration—accuracy and abstraction—took a prominent place among the concerns voiced before the committee. William Ogle, Farr's successor as head of the registrar general's statistical bureau, illustrated his criticism with the example of Norwich, where in 1882 fewer than a quarter of inquests in cases of death from disease specified the pathological cause of death, with the remainder simply returned as "natural."[49] Even when inquest juries tried their hand at medical precision, Ogle suggested, the results were no more illuminating. "One verdict came before me a little while ago," Ogle recalled, "which was this: 'Died from stone in the kidney which stone he swallowed when lying on a gravel path in a state of drunkenness' . . . Another one is this: 'Child, three months old, found dead, but no evidence whether born alive.'"[50] Concluding his testimony on inquests, Ogle drew attention to the incompatible terms of explanation used by inquests and death registration: "I should like to say that it appears to me that the coroners are not carrying out the instructions of the Act. The Act says they are to ascertain the cause of death, and I hold that to call it 'Natural Causes' or 'Visitation of God,' is not ascertaining the cause of death." Encouraged by a committee member's rephrasing of his complaint that such inquest verdicts were "using a phrase instead of giving you a fact?" Ogle concurred through reiteration: "It is using a phrase instead of giving a fact."[51]

The parliamentary committee took disturbing evidence like Ogle's to heart, professing deep concern at "the serious possibilities implied in a system which permits death and burial to take place without the production of satisfactory medical evidence of the cause of death." Thus, despite avowed reservations about mandating the intrusion of official public scrutiny into the most private of occasions—about "piling up the terrors of dying," in the words of its medically qualified chairman—the committee's final report recommended the appointment of medically competent public certifiers to secure the integrity of the nation's vital statistics.[52] "As far as may be," the report observed, "it should be made impossible for any person to disappear from his place in the community without any satisfactory evidence being obtained of the cause of his disappearance."[53]

By insisting on society's right to an account of "disappearances," the committee proposed to make each individual death a public matter. But if this endorsement of a system of expert certifiers seemed to promise a solution to the problem of uncertified deaths, at another level it merely refocused the more fundamental question of precisely how to make death a unit of usable public knowledge. The *British Medical Journal* (*BMJ*) recognized this, remarking that an official certifier "working in his private study" was no match for the powers of informal inquiry and public judgment that an inquest could bring to bear. "The substitution of the certifier's secret report for the coroner's open inquiry in cases where there is 'talk among the neighbours'" would not be an unmixed good, the *BMJ* opined, underscoring its doubts by noting the certifier's "French" pedigree and quoting with approval (and added patriotic embellishment) the committee chairman's assertion that "we English would certainly resist . . . 'the piling up of the terrors of dying' by making so much fuss about it in ordinary cases."[54]

This ambivalent attitude to the ideals of accuracy and openness did not reflect a mere vague inconsistency on the part of medical men. By examining medical views on two basic issues related to death certificates—who should be allowed to see them and who should be allowed to give them—the final section of this chapter will suggest that there were uses to ambivalence. As for the former question, medical spokesmen came increasingly to insist that a modern system of death inquiry was a public good that required a relationship of confidentiality between the practitioner and the state. Inquiry in public produced distortion. On the other hand, the desire for an exclusive right to assist the state in producing an account of public mortality (i.e., the desire for the thoroughly modern ideal of professionalization) paradoxically required these same spokesmen to embrace the archaisms of publicity embedded in the inquest.

III

Medical practitioners were rarely accused of willful deception in certifying individual deaths. There was a notable exception to this rule, however: deaths that might pose a social, financial, or emotional burden on the surviving family—attributable to syphilis, alcohol, and suicide, for instance—were commonly thought to be systematically underreported. The complaint of unreliable returns under such circumstances was an old one. John Graunt, in one of the first English tracts on vital statistics, observed in 1662 that returns for the "French Pox" were artificially low and were classed euphemistically under "ulcers" or "sores." In his analysis of the London Bills of Mortality, Graunt found that in only two districts containing the "vilest and most miserable houses of uncleanliness" was the French Pox reported at all, from which he concluded that "only hated persons, and such, whose very *Noses* were eaten off, were reported by the *searchers* to have died of this too frequent Maladie."[55] Only those powerless to exact a penalty on individual certifiers, in Graunt's view, were subjected to the regime of unvarnished truth.

Graunt's observations on the social basis of distorted returns resonated with nineteenth-century public health modernizers like Henry Rumsey. Inducements for practitioners to conceal or misrepresent causes of death were real, understandable, and dangerous, Rumsey insisted in his 1875 discussion of the errors of death certification. He illustrated his concern with reference to a specific case: A London practitioner of his acquaintance had certified a death that was patently caused by chronic alcoholism as "gastro-enteric disease." When Rumsey questioned his colleague on this vitiation of the nation's vital statistics, the practitioner allegedly replied: "What can I do? It is impossible for me to put the real cause in the certificate;

the family would never forgive me." The danger, for Rumsey, was grave, and the solution clear: "At all events, the medical attendant ought to be released from his false position of responsibility to the friends of the deceased in the performance of his duty to the State."[56]

Rumsey's was among the early voices in what over the next several decades became a rising chorus from government committees, medicolegal societies, and the medical press, urging the introduction of a secret system of death certification as a way of turning doctors into unambiguous agents of accurate mortality statistics. The 1893 Select Committee on Death Certification, acknowledging the weight of testimony urging confidentiality "with a view to removing any inducement to such misrepresentation," recommended, in the interest of statistical accuracy, "that medical practitioners should be required to send certificates of death directly to the registrar, instead of handing them to the representatives of the deceased."[57] The 1904 Interdepartmental Committee on Physical Deterioration came to similar conclusions when it complained that its attempts to gather data on syphilitic diseases had been compromised by systematic underreporting, a fact attributed to the public nature of death certificates.[58] Of the numerous witnesses advocating a system of secret certification as the best remedy for this gap in knowledge, none did so more vigorously than the chairman of the British Medical Association's Medico-Political Committee, Sir Victor Horsley.[59] Under the existing system, Horsley complained, the practitioner's duty to certify death imposed on him two irreconcilable demands: the certificate was at once a document "for the information of the family" and "the scientific basis of statistics for the nation." It was clear to Horsley whose needs took precedence: "Any document certifying the cause of death ought in the opinion of the British Medical Association to be a scientific document, a Government document, what we may call a State paper and a privileged paper, to be given to the Government official, the Registrar General, and the contents of that document should only be communicated to the friends or relatives of the patient at the Registrar General's discretion."[60]

The committee's report, accordingly, added its name to the growing list of those calling for the direct transmittal of certificates from doctors to the GRO.[61] Secret certification was not implemented as a matter of law, however, and those pressing the case were left only with intermediary practices to accomplish their ends. A creative approach to the language of certification was one possible option for the compromised practitioner. If a medical attendant wished to alert the coroner to a suspicious death behind the backs of the family of the deceased, he could issue a cause of death comprising contradictory elements. "It often happens in cases in which the doctor, wishing to keep in with the relatives, will give a certificate which he knows the registrar will not accept," London Coroner H. G. Wyatt explained. "If an inquest is held," Wyatt continued, "they say 'It was not my fault,—it was the coroner who required the inquest.'"[62] The GRO's own Dr. Ogle pointed to another resource: in the absence of provisions for confidentiality, practitioners sometimes certified causes of death in Latin as a means of keeping potentially embarrassing causes of death out of the public eye. He explained that, while registrars would normally refer certificates bearing the words *alcohol poisoning,* for example, to coroners for investigation, they might regard a Latin equivalent as a "naturalizing" certification written in the normal language of medical expertise.[63] Indeed, in the view of another London coroner, the deployment of esoteric nomenclature might be elevated from an informal practice to an officially sanctioned system. Ingelby Oddie, in a letter to Horsley supporting the British Medical Association (BMA) chairman's crusade against the distortions caused by public certification, offered a vision of salvation through obscurantism:

> I am wondering whether practitioners might not be induced to give truthful certificates in cases e.g. of syphilis and alcoholism by introducing terms for death certification which lay relatives would not understand and would not in all probability seek to understand. Take the case of a man of position dying from chronic alcoholism. His doctor would not dream of saying so in his certificate. Why should he not put Hepatic Cirrhosis C_2H_6O—or some similar term? And in Syphilis—the use of the word "specific" so and so would suffice.

These devices, Oddie concluded, would channel the necessary information to the appropriate quarters "without exciting either the suspicion or animosity of the relatives."[64]

Keeping death out of the public eye was in this sense seen as favorable to medical practitioners—secrecy was a condition for successful practice.[65] But in arguing this point, medical spokesmen were able to cite a "public benefit" to privacy alongside the professional one by drawing attention to pathetic example cases like that of Charles Cook. Cook, a London railway clerk, had written a letter to the secretary of the BMA Medico-Political Committee in June 1922 complaining that, on the basis of a blood sample, the Waterloo Hospital had certified the cause of his three-year-old daughter's death as "Congenital Syphilis, Cardiac Fever, and Debility." Cook protested that neither he nor his wife was syphilitic, an assertion supported by negative test results subsequently conducted on each of them. Nevertheless, the public certificate had done its damage: "The net result of losing the child," Cook wrote, "has been the break-up of our home, the wife naturally believing me to be guilty of misconduct with other women, which I am ready to swear upon oath that it is not true."[66] Cook's tale of victimization so perfectly illustrated the profession's case for secret certification that the BMA secretary forwarded it to the GRO. In a personal reply, the registrar general acknowledged the problems posed by such cases, lamenting that "the letter is a very typical instance of the case upon which, as you say, the movement in favour of confidential certification has largely been founded."[67]

At one level, medical representatives expressed and enacted opposition to an overly public English way of managing death. In urging confidentiality, the profession could appeal to a confluence of interests. Confidentiality not only enhanced doctors' own conditions of practice, but also accommodated a dual public interest before death: the need for reliable information at the level of a statist conception of the public and the legitimate desire for privacy on the part of a public comprised of affective individual relationships. But this vision of harmony between medicine and its publics was limited. At the same time as they argued for secret certification to spare the practitioner, the state, and the family from the invidious intrusion of publicity, organizations like the BMA and the Medical Defense Union urged an exemplary visitation of this very form of intrusion in cases of death not supervised by a registered practitioner. An overly public way of managing death, in other words, had its uses in the struggle to secure and enforce professional claims to a monopoly in matters of public mortality.

To show the workings of this alternative vision of medical and public interest, I must return briefly to the mechanics of death certification. In 1845, the GRO began distributing printed "certificates of the cause of death," intended for the sole use of licensed practitioners.[68] Registrars were instructed to record a death as "certified" only on the basis of information from such a practitioner, preferably entered on the printed form. In the absence of this documentation, the registrar, if satisfied that the unlicensed source of information was trustworthy, could register the death on his own authority, but only as an "uncertified" death—not, that is, determined by a recognized medical source. Alternatively, he could refer the death to the coroner for further scrutiny. By establishing a hierarchy of sorts between recognized medical practitioners authorized to give certified causes of death and the "uncertified" information of unlicensed informants, the GRO took a first step toward recognizing licensed medical practitioners as the privileged source for the civil registration of death. It established a classificatory means for distinguishing between sources of information, allowing proponents of medicalized registration to identify unsatisfactory causes with the normatively charged designation "uncertified."

Yet the introduction of this distinction was not accompanied by a directive to registrars to refuse registration merely on the grounds of it being "uncertified." This gap was supposed to have been filled with the passage of the 1874 Birth and Death Registration Act, which restricted certification rights to those practitioners recognized under the terms of the 1858 Medical Act, and imposed upon registered practitioners a statutory (but unremunerated) obligation to provide certificates of death to registrars for all deaths that took place under their care. These developments were seen by the established medical community as a decidedly mixed offering, however. Though the 1874 act ostensibly compensated the imposition of a statutory duty on registered practitioners to certify by reinforcing their exclusive right to produce a certified cause of death, unregistered practitioners could still participate in the bureaucratic management of death by having their information registered (as "uncertified") without necessary recourse to orthodox medical opinion.

Representatives of the established medical community responded to the act with a campaign directed specifically at the dangers of allowing unlicensed practitioners to take part in the death registration system. In February 1876, a *Lancet* editorial commented on the response made by the secretary of state to a parliamentary question about the acceptance by a Staffordshire registrar of an inordinate number of certificates from unlicensed practitioners (nearly half the total of registered deaths, according to the questioner). To the secretary of state's laconic reply that nothing but "information" was taken from this assortment of unorthodox practitioners—and not a *certified* cause—the *Lancet* retorted: "This is a very cool answer to make to a Member of Parliament who is jealous of the value of public records. That nothing but *information* of the cause of death is to be taken from bone-setters and herbalists! Why, this is the very thing they cannot give." The very viability of mortality statistics was at stake, the editors concluded, because, for death certification to have any worth, "it must represent medical knowledge and not the absence of it."[69]

But a disinterested concern for the sanctity of the nation's vital statistics was not the sole motive of the medical profession, as the *Lancet*'s concluding observation made clear: "No wonder that the profession asks what is the advantage of being registered."[70] Admitting causes of death attested to by unregistered practitioners, even if only as "uncertified," disrupted both mortality statistics *and* medical professionalism. Incorporation of indiscriminately derived causes into the nation's ledgers of death, according to the *Lancet*, effectively encouraged the public's resort to unorthodox treatment by making no practical distinction between medical regimes of monitoring death: "The deluded people find the certificate of the quack as effective for registration as that of the qualified and registered practitioner."[71]

This problem stimulated several reform proposals, among which was the suggestion that all deaths not certified by a registered practitioner be referred for investigation by the coroner. On the surface, this was a straightforward and relatively innocuous innovation; notification to the coroner would not necessarily lead to a full inquest, and a coroner's refusal to hold an inquiry upon such notification would again lead to an "uncertified" entry into the mortality statistics. But what, for advocates of medical professionalization, was most appealing—and paradoxical—about recourse to the inquest was the corrective power entailed in threatening to visit the terrors of an intrusive public inquiry upon a public unwilling to recognize the self-evident value of having orthodox medicine at its deathbed.

The workings of this tacit visitation of exemplary discipline were made the subject of extended discussion for members of the Coroners' Society around the turn of the century. In 1897, the society's secretary, A. Braxton Hicks, wrote to the Home Office urging that cases of death under unlicensed care be referred as a matter of course to the coroner. In the absence of such a provision, Hicks warned, the public was endangered by unauthorized practitioners who "are enabled to carry on their ignorant, unlawful, and unscrupulous practices, and openly boast they can give a certificate and no inquiry will be made."[72] Referral would serve several purposes at once: it would provide a way of investigating the circumstances of individual cases; it would enable coroners to educate the public about the shortcomings of unregistered care; and, by reliably associating uncertified death with inquests, the practical consequences of ignoring registered medical care would in and of themselves induce a salutary change.

Hicks's successor, Walter Schroeder, illustrated the potential value of this practice by directing the attention of the society to a district where uncertified deaths had been commonplace and where proper medical advice had rarely been sought before death. When the local coroner instituted a new policy of holding an inquest in these cases, the results were impressive: "In a very short time the inhabitants of the district saw the advisability of early medical advice, and I doubt very much if there is now ever an uncertified death (except it is of an occasional prematurely born child, or a very old person) in that neighbourhood, and inquests, I know, are very rare."[73] Instances in which the threat of inquests had been instrumental in rooting out quackery were also publicized in the medical press. Birkenhead's medical coroner was singled out for praise by the *BMJ*, for instance, when in 1902 he declared his intention to hold an inquest in every case of death not certified by a registered practitioner. He had taken this step, he explained, in the hope that "inquests held in such cases as these would serve to show the public the stupidity and futility of seeking medical advice from people who were not qualified to give it."[74]

The desire to educate the public out of its "futile stupidity" revealed a contingent attitude of the medical profession to the "immoderation" of public inquiry. Publicity could at the same time lead to both divided loyalty and error and serve as a tool of enlightenment. It was the duty of the medical practitioner to shield the public from a grotesque intrusion in the former case and to mobilize intrusion in the service of reform in the latter. It was these very excesses—specular, emotive, intrusive—that held the promise of modifying public behavior by visiting upon friends and relatives an unwelcome and quasi-punitive procedure interrupting the smooth transition from death to burial. By insisting upon full public inquiry into these types of death, promoters of the cause of medical professionalism embraced this otherwise reviled feature of the inquest system as a way of denying recalcitrant deaths the benefits of a rational, efficient, and scientifically purified encoding.

This process of education, of course, required coordinated action on the part of registrars and coroners. A 1914 directive from the registrar general attempted to secure exactly this by instructing his subordinates that "Any Death for which no Certificate of Cause of Death is produced from a Registered Medical Practitioner must be reported by the Registrar to the Coroner if this has not been done already."[75] Yet the value of this measure still depended almost entirely upon the attitude adopted by local coroners to the cases reported to them. In many such instances coroners either simply returned a "No Inquest" form to registrars or neglected to respond at all, and in these jurisdictions the amended rules quickly fell into desuetude. However, the promise of an unbounded public inquest system allied to scientific rationalism (in this case standardized medical care) was amply demonstrated in other coroners' districts.

Consider, in conclusion, the only cases of protest against the 1914 directive that survive in the official record, which involve two

Northern herbalists and the national office of the Christian Science Church.[76] In 1925, the registrar general met with W. Burns Lingard and Korah Culpan, Yorkshire herbalists complaining of persecution at the hands of their local coroner. Culpan presented evidence of the expressions of public outrage that followed the local coroner's decision to hold inquests in all deaths of his patients. Protest meetings had been held, and a petition had been circulated attesting to the "very strong resentment that bereaved families should suffer the painful, additional and undignified burden of what we are sure are uncalled for Inquests." Culpan informed the registrar general that, as a result of the coroner's actions, he had instituted a practice of asking the family of a terminal case whether they wished to have a registered practitioner called in for the purposes of certification. In nine of ten cases, he claimed, the families refused, preferring to "risk the humiliation of inquests, etc." Lingard added to Culpan's complaints by describing a meeting between himself and his local coroner, in which the latter dismissed the herbalist's protests that the deaths under his care were as natural as any under a registered practitioner's by observing: "You know I could regard the mere fact that you attended a deceased person as a suspicious circumstance."[77]

Soon after the herbalists' hearing, another Whitehall office received a similar set of complaints. Charles Tennant, district manager of the Christian Science Committee on Publications for Great Britain and Ireland, wrote a lengthy letter to the lord chancellor detailing the despotic practices of many English coroners. Again, the charge was the unjustifiable and systematic holding of inquests in patently "natural" cases merely because of an absence of registered medical care. In such cases, Tennant wrote, coroners order unnecessary examinations, go into the case history in searching detail, and at the end of the hearings call the surviving relative to the witness box "and describe to him or her what the post-mortem examination is said to have revealed as to the cause of death, and, as already stated, sometimes adding his personal disapprobation and even ridicule of the conduct of the deceased and the relatives."[78] Tennant was also very clear on the emotive chord struck by these instances of inquisitorial excess: "Your Lordship will readily appreciate the horror," he advised the lord chancellor, "with which most people shrink from the idea of a post-mortem examination upon the remains of one they loved, and when the gruesomeness of such an examination is heightened by the compulsory removal of the body to a public mortuary, it is easy to imagine the added sufferings endured by those who are already in distress."[79]

The use of the inquest as a public stage for the ritual humiliation of those refusing to submit to the authority of orthodox medicine, then, was secured by its treatment of the body. But this was only one instance in which the tension between the body as scientific and as public object proved significant for the making of the modern inquest. This is not surprising: The body, as the inquest's jurisdictional, symbolic, and epistemological center, was necessarily a pivotal feature of any vision for thoroughgoing reform. The following two chapters are devoted to a more sustained consideration of the implications of this simple fact.

FROM THE ALEHOUSE
TO THE COURTHOUSE

Bodies and the Recasting of Inquest Practice

In his 1889 *Life and Times of Thomas Wakley,* medical journalist S. Squire Sprigge cast his biographical subject as the savior of the modern inquest. Before Wakley, Sprigge declared, the inquest had been on the slippery path toward well-deserved oblivion. He introduced his readers to his chapter on "The Old-Fashioned Coroner's Inquest" by way of a striking depiction:

> The tint of the tavern-parlour vitiated the evidence, ruined the discretion of the jurors, and detracted from the dignity of the coroner. The solemnity of the occasion was too generally lightened by alcohol or entirely nullified by the incompetency of the judge. In short, the tribunal designed by Edward I. to be one of the most important in his kingdom . . . had been universally degraded to a dreary farce, . . . a thing proverbially to be laughed at, and where the majesty of death evaporated with the fumes from the gin of the jury.[1]

In this passage Sprigge drew attention to what for modern readers is surely the most curious feature of nineteenth-century coroners' courts: namely, that there were no courts; that, in place of an officially designated site from which to conduct the business of inquests, most inquiries into the unexpected deaths of English men and women took place in public houses.

In Sprigge's portrait of Wakley as an indefatigable reformer before whom archaisms and anomalies were swept away with dispatch, pub inquests stood as emblematic of precisely what Wakley had successfully worked to overcome. Yet Sprigge is wrong on both counts. First, at the time of his writing, the pub inquest was by no means a distant memory. In London the practice had only begun to decline significantly in the last decade of the century, the result of an express policy on the part of the newly constituted London County Council (LCC), bolstered by legislation specifically aimed at the metropolis, to wipe out inquests "held amid the desecrating surroundings of the public house."[2] Other provincial authorities followed London's attack on "desecrating" sites of inquiry, spurred on by the 1902 Licensing Act, which forbade pub inquests if any reasonable alternative existed. But changes were slow and partial. Olive Anderson notes, for instance, that Lancashire inquests were conducted in pubs in virtually every case at least as late as the 1890s.[3] Indeed, as a Home Office survey taken in the early 1930s discovered, pub inquests, especially in rural districts, survived well into the twentieth century.[4]

Second, Sprigge's hagiographic assumptions notwithstanding, Wakley had had little if anything to say about pub inquests. This is not entirely surprising. As Brian Harrison has pointed out, the early nineteenth-century "public house"—as one of the only buildings to which every denizen of a village, town, or city had access—was regularly pressed into service for a wide range of civic needs: "Coroners held their inquests there, doctors interviewed their patients there, governments collected their taxes there, the authorities held their prisoners there, and publicans even sometimes acted as registrars of births and deaths."[5] As a creature of the tavern-centered politics of London radicalism, furthermore, Wakley was no stranger to the multiple uses of these important civic sites. For him, inquests at pubs might well represent not merely an unremarkable reality of civic life, but an expression of a properly functioning popular politics.[6]

But Sprigge is undoubtedly correct in registering a late nineteenth-century mood of impatience, annoyance, and embarrassment about the connection between medicolegal inquiry into death and public houses and about the way that this association seemed at once to represent and to reinforce the inquest as a singularly profane institution of English civil life.[7] From the final quarter of the nineteenth century, the pub-based inquest served as a kind of rallying point for a campaign to transform the spatial and conceptual grounds for public inquiry into death, a campaign that sought to establish a new set of boundaries: between the lay person and the medical expert; the dead and the living; the purposeful and the prurient; the sentimental and the instrumental.

In this process of boundary drawing, the physical corpse—as the symbolic, evidentiary, and operational heart of the inquest—took center stage. Critiques of the inquest's regime of the body focused on three of its underlying features: first, at the inquest's basic physical arrangements; second, at the modes of interaction between the lay public and the dead body; and third, at the provisions for expert inquiry which seemed to hinder an accurate reading of the body in question. This chapter concentrates on the first two items on the reformist agenda, demonstrating the significance of the rhetorical and material restructuring of the boundaries that set the terms of engagement between the inquest and the body as its prime object of contemplation. Chapter 4 builds on this analysis, locating the tension between the profane and the scientific within conceptions of medical evidence itself, in its terms of derivation, significance, and circulation.

This line of inquiry shifts the focus away from an obvious "politics" of the inquest and onto a more technical, administrative, and jurisdictional plane. From one perspective, this shift might be interpreted as a clear victory for "modernizers." That is, as the terrain of debate moves from the hustings and political meetings to the representatives and institutions of an increasingly organized state medicine, the peculiar expert-popular hybrid that shaped discussions about the early nineteenth-century inquest withers away.

There is some truth in this reading: after Wakley and his ilk, there is a clearly a less radical (and radicalizing) edge to the hybrid.[8] But this should not blind us to the fact that the project of medical reform continued in significant ways to speak the language of popular liberties and to speak it in a manner similar to the heyday of radical medicine—that is, partially, contingently, and strategically. Therefore, even as the coming chapters represent a break in the historical materials with which they engage, they probe new areas with a recognizable set of analytical tools, tools designed to uncover the instabilities of the medical reformist agenda even as it becomes ostensibly more self-assured. At the same time that they were trumpeting the case for an inquest cast in the image of science, this analysis indicates, architects of reform were deeply concerned about the possible shortcomings of a fully medicalized format for inquiry. Although the pub inquest was disparaged as the exemplary sign of inefficiency, disorder, and irrational archaism, the frustrations to scientific inquiry that it represented were also recognized as springing from important—and in some senses necessary—political exigencies. Medicalizers at once constructed the unreformed inquest as a barbaric other and attempted to rechannel the elements of access and publicity through an acceptable filter of scientific mediation.

I

The pub inquest had drawn the attention, favorable and unfavorable, of commentators well before Sprigge. Charles Dickens figured among its more notable midcentury critics. Both in novelistic satire (most famously with *Bleak House*'s riotous "inkwich" held at the Sol's Arms tavern) and somewhat more systematically in his journalistic writings, he depicted scenes of degraded inquest proceedings.[9] "A Coroner's Inquest," published in one of the early numbers of Dickens's popular weekly, *Household Words,* began with the following lament: "If there appeared a paragraph in the newspapers, stating that her Majesty's representative, the Lord Chief Justice of the Queen's Bench, had held a solemn Court in the parlour of the 'Elephant and Tooth-pick,' the reader would rightly conceive that the Crown and dignity of our Sovereign Lady had suffered some derogation. Yet an equal abasement daily takes place without exciting especial wonder."[10] Coroners, Dickens explained, in theory "no less delegates of Royalty" than the lord chief justice (himself officially designated "first coroner of the land"), were regularly obliged to conduct their inquiries in public houses, "amidst several implements of conviviality, the odour of gin and the smell of tobacco-smoke."[11]

But interventions such as this did not constitute a widespread or organized program for change. In fact, the physical environs of inquests played no discernible part in the initial decades of medically inspired reform, and it was not until the final quarter of the century that the pub inquest became an object of sustained scrutiny. Drawing his immediate inspiration from the press's excoriation of the

Representation of a tavern inquest at midcentury. Thomas Wakley is pictured in the center, with Charles Dickens to his left. Courtesy, Bodleian Library, University of Oxford.

notorious 1876 inquest on Charles Bravo, which depicted the hearings as a crass indulgence in gossip mongering all too appropriate for its base setting, Lord Francis Hervey took to the floor of the Commons to call for legislative overhaul of inquest practice. Amid a litany of complaints, Hervey demanded of his parliamentary colleagues: "Could anything be more prejudicial to the proper holding of a Coroner's inquiry than the holding of it, as was so often the case, in a public-house? Surely there was something perfectly disgusting in holding an inquiry so solemn and sometimes so delicate in a public-house with jingling glasses and the shouts of drunken persons all around."[12]

If the immediate context for Hervey's motion was a particularly well-publicized instance of inquisitorial intemperance, there is no question that its power derived from contemporary trends of a more fundamental sort. Most broadly, his denunciation of the pub inquest can be seen as a sign of a deepening anxiety about traditional bastions of popular politics in the era of mass democracy and the consequent attempt to construct a more disciplined and decorous citizenry. As John Belchem and James Epstein have recently observed, the reordering of public (especially urban) spaces was a crucial component of the mid-Victorian attempt to forge this "civic vision based on a trans-class vision of respectable citizenship."[13]

Temperance and public health provided two further and more obvious sources of inspiration. As Hervey's intervention suggests, the coupling of the inquest and the tavern easily lent itself to the literal charge that drink clouded judgment, and in this he was by no means alone. The *British Medical Journal* (*BMJ*) took the opportunity of informing its readers of the frequency of pub inquests in Liverpool (which it reported as nearly 95 percent in 1877) to denounce the practice as a threat to the value of evidence "in consequence of the facility with which the witnesses can obtain drink."[14] John Payne, coroner for the City of London and Southwark, warned in the same year of the dangers of inquests conducted through the veil of alcohol, observing that pubs afforded "a great inducement to witnesses to take more than they ought."[15]

Reflective of a more diffuse, temperance-inspired critique of tavern culture, the pub inquest also functioned in reformist discussions as an emblem of the inquest's intrusive, prurient nature, by which it seemed to merge with its uncontrolled surroundings. Petitions submitted to the Home Office by the borough councils of Sheffield, Kingston upon Hull, and Halifax in 1889 traded heavily on this environmental critique. Removing inquests from the "unwholesome surroundings of a room in a Public House to a Court House," they insisted, would both "promote the dignity of the inquiry and lead to increased efficiency." The petitioners then contrasted the promised dignity and efficiency with contemporary practice: "Nothing can be more repugnant to bereaved relatives than to have their natural sufferings accentuated by being the objects of what often proves to be unjust suspicions, exhibited to the gaze of persons surrounding a Public House bar."[16] The physical context of the inquest, *The Times* agreed, was pivotal for any attempt at purification. That "the inquiries into the tragedies of life are often conducted in the squalid atmosphere of third rate taverns," an editorial insisted in October 1895, was among the longstanding complaints made by the public against the inquest, constituting nothing less than "an excrescence

and an anachronism." The editorial concluded with a plea for a reformist vision that would extend beyond the timeworn and "fringe"
debate about legal versus medical coroners and would be capable of
sparing the nation "that mixture of ghastliness and low conviviality
which sometimes surrounds the proceedings when held in busy public houses."[17]

Sentient participants were not the only ones seen as affronted or
compromised by the context of inquest proceedings. The body, too,
shared the space of the tavern, and it was here that sanitarians entered the fray. Prior to the concerted exercise in municipal mortuary
construction that commenced in the last quarter of the century,
bodies coming under the scrutiny of an inquest lay in one of several
places awaiting examination. Deaths occurring in places already
equipped with a deadhouse of some kind (hospitals or workhouse
infirmaries, for instance) were generally dealt with on the premises.
Keeping the body where it lay was also possible when deaths occurred in private homes, though these were often subject to removal
on a coroner's order. Bodies discovered outdoors were commonly
transported to the shed of the nearest tavern. Indeed, such was the
supposed ease with which pub buildings were appropriable for this
purpose that the Licensed Victualler, the publicans' trade journal,
considered it necessary to include among its trade advisories the following entry: "Dead Body. In some quarters there is a belief that an
innkeeper is compelled by law to receive into his house a dead body,
found in the streets or washed ashore, for the purpose of an inquest,
but such is perfectly erroneous. Inns are established to supply
the wants of the living, and have nothing to do with the dead."[18]
The Victualler's admonishment squared with contemporary judicial
opinion: in 1894, a South Staffordshire publican sued the local constabulary for loss of trade that he attributed to the noxious fumes
emanating from a decomposing body left for days by the police in
his pub's club room. The judge upheld the suit, remarking in his
summation that "some people imagined (erroneously) that because
a public house was open to every member of the public lawfully fre-
quenting it, a dead body could be taken there."[19]

The 1875 Public Health Act drove an important statutory wedge
between the tavern and the dead body by including a provision

enabling the construction of public mortuaries and coroners' courts by local authorities out of public funds.[20] In its wake, medical and public health journals began monitoring the issue in earnest, announcing with due fanfare any plans for new construction and chastising delinquent local authorities for endangering the health and morals of their constituents. The *Sanitary Record,* one of the movement's leading organs, greeted the November 1876 opening of the Westminster complex in typical fashion, hailing it as a victory for civic-minded sensibility and voicing its hope that soon all districts would boast "one of these indispensable aids to health and decency."[21]

"Decency" was clearly on the minds of the City of London Corporation when it met in January 1877 to approve plans for a spatially reconstituted inquest. The corporation's commissioners of sewers resolved to put an end to inquest practices performed "to the great disgust of every person concerned, and the discredit to our great City, which ought to provide the necessary requirements of decency and decorum in the administration of this important branch of the law."[22] At the cost of twelve thousand pounds, the commissioners provided the district with a model inquest site, featuring a coroner's court, a deadhouse fitted for postmortem examinations, a laboratory with weighing and consulting rooms, a keeper's house, and sheds for a disinfecting apparatus and an ambulance. The city's coroner was fulsome in his expression of gratitude for the corporation's lifeline to the respectability and salubrity of his ancient office: "I rejoice to think that after 30 years' experience in holding inquests in public houses, I shall now, through the great kindness of the Corporation of London, have a Court where the inquiry can be conducted with quietude and propriety."[23] The *Lancet,* for its part, touted the corporation's achievement, celebrating it as an ideal inquest complex that substituted the "grotesque surroundings" of the tavern and the outhouse with one "in an isolated area, replete with every convenience for the due performance of the coroner's function."[24]

This juxtaposition of the "grotesque surroundings" of the pub with the "convenient" and "isolated" mortuary became a touchstone for the idealized space of a self-contained inquest. Indeed, the necessity of insulating the complex for managing death, physically and conceptually, from the outside world was a design imperative written into recommendations for the appearance, material composition, and location of the structures themselves. Mortuaries and courtrooms, the London County Council's architectural plans stipulated, should strive for solidity and distinctiveness, a permanence and fixedness of place so clearly missing in the case of pub inquests: "The buildings should be substantial structures of brick or stone. In their external appearance attention should be paid to such architectural features as may serve to convey the impression of due respect for the dead."[25]

"Respect for the dead" could also be inscribed in practical and aesthetic detail. Paddington's new inquest site, built in 1902 as one of the last in this wave of construction, was described in the local press as conveying this sensibility in a hybrid idiom of Christian piety, sanitation, and science. It contained not only a postmortem room but also a microscopy room, which put it at the forefront of modern techniques of death investigation. "The design," the article continued, "is a pretty one, not the least effective portion of the building being the Cathedral entrance porch which will be surmounted by two Maltese crosses in white stone." The article concluded in praise of the "special attention paid to ventilation, lighting, and sanitary arrangements."[26]

The interior layout of an ideal court-mortuary complex, moreover, enforced an orderly flow of activity within, emphasizing containment and separation. The courtroom itself was modeled on other legal spaces, with a judge's dais at the head for the sitting coroner, a jurors' dock, and rows of benches for witnesses, the public, and the press. The body whose story was to be unfolded lay in proximate isolation, accessible yet decently insulated behind the deliberating chamber. The LCC's model plans were equally concerned with the relationship of the mortuary to its external environment, recommending that sites should be at least five thousand square feet in dimension and "should, as far as practicable, be free and open, not surrounded by lofty buildings or in close proximity to dwellings."[27]

Yet internal and external separation were often difficult to realize in practice. Holborn's mortuary presented a clear instance of failed internal regulation, "an ill-designed building in a confined situation, containing three rooms, viz. a mortuary, post-mortem room and a small infectious chamber which also serves as a viewing room to the mortuary. The approach to the mortuary chamber is either through the post-mortem room or infectious chamber."[28] Such insufficient provision for internal order risked the reintroduction of the unregulated play of encounters between the living and the dead that had made the pub inquest an outrage to decency. Furthermore, even the best vestry efforts to purchase land for the erection of mortuaries and courts could run up against local concerns that the mortuary would not be adequately isolated from its surroundings. When the Camberwell vestry entered into negotiations for the purchase of a piece of property, the landowner's agent reminded the vestry that the sale would "undoubtedly injuriously effect the neighbouring portion of the estate."[29] The terms of sale for the land slated for the Hammersmith complex stipulated the erection "of a wall at a reasonable height to shield the property from the adjoining land," the seller noting that "a coroner's court is not a desirable neighbour."[30]

Inadequate provision for separation led to scandal. Under the headline "The Condition of Bermondsey's Public Mortuary," the *Southern Recorder* of 12 January 1889 reported on a controversy swirling around one South London site: "Public attention has recently been called to the alleged insanitary and semi-public condition of the mortuary owned by the parish of Bermondsey." The *Recorder*'s framing of the issue is noteworthy for its repeated invocation of the mortuary's "public" status—three times in the headline and lead sentence alone. The mortuary (public because funded out of local rates) was presented to the public as inadequate to its public mandate because, being "semi-public," it had failed to enforce a regime of decorous separation between the living public and its dead. For those members of the public not aware of the mortuary's overly public location, the *Recorder* obligingly gave its coordinates: "It may be added for general information that the mortuary is a small one-storey structure erected beside the parish churchyard, at its eastern boundary. Many might pass and re-pass it without being aware of its existence."

The insinuation of the civic undead as an unnoticed feature of the everyday landscape, it seemed, posed a danger, both because even

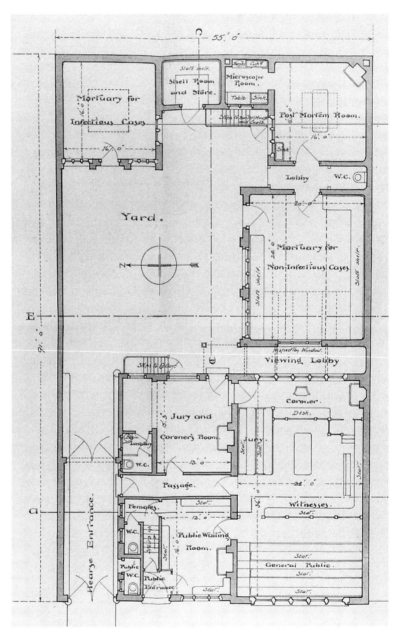

LCC-Suggested Model Plans for a Coroner's Court and Mortuary, 1893.
Courtesy, London Metropolitan Archives.

Courtroom at Paddington Coroner's Court, 1910. Courtesy, London Metropolitan Archives.

unconscious encounters had deleterious sanitary consequences and because the places where they lay pending final communal judgment could not remain unnoticed forever. Detection was inevitable, occasioned by the sighting of an exposed corpse in transit to the mortuary, most likely, or by the discovery of a miasmatic stench issuing from its chambers. By that time, however, a demoralizing familiarity might have already set in. Conflating the imperatives of health and morals was central to this conception of insularity; as one parish officer wrote, the present situation of the mortuary was "detrimental to the morals of the people," observing that the building failed in its two main requirements—that it "should possess every sanitary requirement, and should also be private."[31]

In the isolated and purposeful space of the court-mortuary complex, an alternative to Sprigge's gin-soaked farce emerged. But even if partially enacted in the ideal world of architects' drawings and newly completed shells of buildings, the desire for regulated contact between the living and the dead, so vividly drawn in reformers' condemnations of pub inquests, was destabilized by one of the central rituals of the inquest, the view of the body.

II

The view of the body was, in a strict procedural sense, just that.[32] After the jurors had been sworn and generally before any testimony had been heard, they retired to wherever the body was lying and, in the company of the coroner, examined it. From the earliest legal writings on inquests right up to the eve of its abolition in 1926, the view by both coroner and jury was considered an inviolate feature of inquest procedure. Early modern commentaries on inquest law indicate that the body was continuously present at its inquest. Sir John Jervis, author of the manual on coroners' law that is used to this day, noted in his first edition (1829) that "it would seem that anciently the body was lying before the jury and Coroner during the whole evidence."[33] This was no longer the case by the period in which Jervis was writing, but the importance of the view as a component of inquests was undiminished by the retreat of the physical body: "An inquest of death can be taken by a Coroner *super visum corporis* only; and if there be no view, the inquisition is void . . . Both the Coroner and the jury must view the body at the same time, for the inquisition proceeds on the view of the body lying dead."[34] Textbooks on inquests from the early part of the twentieth century confirm their nineteenth-century predecessors on this score, as did the Lord Chancellor's Office when, in 1898, it declared the view "essential, not only in law (it is the view of the body that gives jurisdiction) but in good sense."[35] Indeed, only a few months before the radical restriction of the view by the 1926 Coroners (Amendment) Act, the High Court ruled that an inquest verdict was invalid because the coroner had neglected to follow proper viewing protocol.[36]

Though a settled feature of inquest law, questions about the practical value of the view featured prominently in the attacks on the unreformed inquest. For its critics, the view of the body by the coroner and, more especially, the lay jury represented the epistemological corollary of the pub inquest. As the pub distracted from purposeful evaluation of causes of death, introducing extraneous, salacious, and damaging play into what ought to have been solemn medicolegal deliberation, so, too, did the jury's view interpose a veil between scientific accuracy and the corpse as scientific object. Arguments for eliminating the view, consequently, shared many terms of reference with earlier criticisms of pub inquests and outhouse postmortems but took the critique to a new level by promising a more radical program of abstraction and insularity, involving at its limits the complete disengagement of the body from the inquest as a popular practice.

In its critics' eyes, the view of the body was an intrusive outrage, at once a residual barbarity out of place in the modern world, a sanitarian's nightmare, and the source of profane interference with the efficient and purposeful production of scientific knowledge. It was, in the opinion of Charles Rothera, the coroner for Nottingham and one of the view's most vocal critics among the members of the Coroners' Society, "useless, revolting, and the cause of illness to Jurors."[37] *Revolting* was a term commonly coupled with the view, which could be deployed in several directions at the same time. The view, for one thing, was patently revolting to jurors, who were forced to enact what many held to be a strange ritual of a less genteel age. Abolishing the view could thus be justified on grounds of taste alone, according to the member of Parliament for Yorkshire, J. S. Higham, being "in harmony with the whole tendency of modern times, because the compulsory viewing of a body was almost on a level with the idea of public executions and other barbarous methods in vogue centuries ago."[38]

While Higham's reference to the abolition of public hanging may seem yet another affirmation of that ineluctable historical trajectory most commonly associated with Norbert Elias's notion of "the civilizing process," a further reading equally recommends itself. Taking as its conceptual guideline the double-edged vision of civilization offered by Freud, this analysis might view refinement as a redistributive rather than a linear and progressive phenomenon, emphasizing the negotiated and indeed compensatory logic of the drive toward a "modern" and "civilized" sensibility.[39] Higham's intervention is apposite in this regard, since at the same time that the legislature abolished direct public participation in the rite of the scaffold, it made an inquest on every executed body a statutory requirement. The inquest, therefore, was explicitly inserted as a proxy—and stood as an implicit indicator of the continued need—for some public relationship to the stuff of execution.[40]

The enactment of the view not only pressed the jury to revolting excess, but also subjected the body and its mourners to the desecration that followed from involvement with the unbounded profane. The view subjected the body to an "objectionable" and "useless" exposure and the relatives to "an intrusion into a home rendered sacred by extraordinary grief and trouble."[41] The dangers attributed to a promiscuous trafficking between the dead and an untrained public shifted easily from those affecting public sensibility to those affecting public health. From the standpoint of sanitarian logic, exposing laymen to the corpse, even if framed within established ritual, invited infection. The pathogenic potential of the view was argued as a straightforward epidemiological truism: Corpses, especially those inhabited by the traces of infectious diseases, threatened to visit their fate upon all who would approach unprotected.[42]

By the turn of the century, the effects of these warnings were being felt both directly and indirectly, as a review of administrative records indicates. Coroners, for instance, began to inform the Home Office of jurors' resentment of the view, and on more than one occasion inquest jurors, convinced that they had contracted a disease from viewing a noxious corpse, applied to the government for compensation.[43] The problems associated with the view also became a matter of commercial notice, as local authorities responsible for providing inquest facilities received announcements of technological breakthroughs from engineering firms anxious to land a municipal contract. One such flier sent to the London County Council heralded the arrival of a revolutionary mortuary container, built to the following specifications: "All Steel. Sanitary. Indestructible. Air Tight. Fluid Tight."[44] The LCC was being handed, in effect, the promise of a mortuary within the mortuary. In another design submission, the maker of the "De Rechter Body Preserving Apparatus" placed at the top of its list of practical applications of the apparatus its benefits to coroners and their juries:

Body Preserving Apparatus

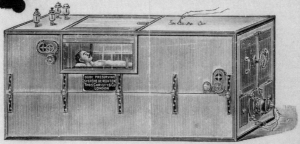

WE desire to call the attention of Coroners and Public Health Authorities to an invention which is capable of rendering incalculable service to all those whose duties bring them in contact with dead bodies. The apparatus—System de Rechter—consists of a box in the form of a rectangular prism which is divided into two parts by metallic shutters.

One is a disinfecting chamber in which the dead body rests on a metal litter of wide meshed gauze. The top and sides are furnished with double glass windows which render the whole of the interior readily accessible to view.

The other is an evaporation chamber containing linen cloths. These cloths are kept constantly moist with Formaldehyde—which is supplied from feeders situated on the outside. The Formaldehyde vapour is then propelled from one chamber to the other by means of a small electric motor, and so long as the motor is working the vapour is kept in constant motion.

The object of the apparatus is the complete disinfection and permanent preservation of any body which is subjected to the Formaldehyde vapour for a sufficient length of time. That it fulfils this object completely we are ready at any time to demonstrate. The fact has already been established in Belgium, where the apparatus is employed in many institutions and can with permission be seen working at the City of London Mortuary, Golden Lane, where it has been installed after investigation on the Continent by the City Authorities.

1. CORONERS AND THEIR JURIES. At present the practice of viewing the body is liable to be not only a particularly trying discipline, but in some cases a very dangerous one. Emanations from decomposing corpses are not only revolting to the senses; they are often exceedingly dangerous to health. A cadaver which is subjected to the influence of this apparatus, no matter how far in decomposition it may originally have been, becomes entirely odourless and perfectly aseptic. Putrefaction is not only arrested; it is absolutely and finally abolished. The process does not in any way interfere with the detection of mineral poisons in medico-legal cases.

2. PUBLIC HEALTH AUTHORITIES. Mortuaries need no longer be the receptacles for decomposing animal matter which, in spite of all precautions, they necessarily are at present, the apparatus being absolutely airtight.

In London the initial cost is **£194**, installed ready for working. Other Cities specially quoted for. With ordinary care the cost of upkeep resolves itself into the inconsiderable items of the necessary electricity and the necessary Formaldehyde.

SOLE AGENTS:—

THOS. CHRISTY & Co., Old Swan Lane, London, E.C.

The De Rechter Body-Preserving Apparatus. Courtesy, London Metropolitan Archives.

> At present the practice of viewing the body is liable to be not only a par-
> ticularly trying discipline, but in some cases a very dangerous one. Ema-
> nations from decomposing corpses are not only revolting to the senses;
> they are often exceedingly dangerous to health. A cadaver which is sub-
> jected to the influence of this apparatus, no matter how far in decom-
> position it may originally have been, becomes entirely odourless and
> perfectly aseptic. Putrefaction is not only arrested; it is absolutely and
> finally abolished.[45]

These technological innovations promised, in theory, a new type of
inquest and, indeed, a new body—one contained, preserved, and
inviolate.

In practice, however, the methods adopted to protect the duty-
bound public from the dangers of postmortem exhalation were of a
more low-tech variety, involving either glass tops fitted over coffins
or glass screens between the body and the jurors when they came to
view. Yet sanitary filters did nothing to placate the view's critics, instead serving to reinforce their third and most telling line of attack.
By mediating between the jury and the body, the modern view's concessions to sanitary considerations had rendered the procedure even
more irrelevant, explained the *BMJ* in 1898: "The body is usually
lying in the coffin or covered with a sheet, with only the face visible.
In all well-equipped mortuaries recently built, the jury actually 'view'
the body through a window, themselves standing outside. How can
such a casual glance help the jury to ascertain the cause of death?"[46]
Mediating screens between the body and the jury might reduce one
form of infection but did nothing except perhaps exacerbate another
scourge visited upon the inquest by the view, that of unscientific error.

The view was not merely an anachronistic barbarism, then, but
also (and more importantly) an epistemological one. In arguing the
latter version of an unwholesome mixing of the living and the dead,
opponents of the jury's physical encounter with the body deployed
a historicizing critique based on progressivist accounts of medical
knowledge and of modes of modern death. On the one hand, they
pointed to advances in forensic medicine and the science of pathology, which had created a new form of knowledge representing an
improvement on anything a lay jury might be able to offer the inves-

tigation. The jury view from this perspective became a relic of an age of egalitarian knowledge that no longer obtained.

On the other hand, critics maintained that changes in the nature of human violence had rendered the contemporary corpse physically distinct from its predecessors and that these changes had made the jury view even more marginal. This historical typology of dead bodies resonated with earlier meditations on inquest work. "As civilization increases the refinement in crime keeps pace," the Middlesex coroner William Baker had observed in his 1851 handbook on inquest law, and this elementary fact had important implications for his office: "In the ruder ages, the means resorted to, to gratify a deep lodged hatred, or to possess property of others was always of a bold and violent description, and left its traces behind, but now villainy is so refined, and so many means have been discovered whereby life may be taken, and the murderer leaves scarcely a clue to his discovery."[47] With the shift from violence against a body that left legible traces affording a surface reconstruction to an inward violation, invisible to the surface gaze of the observer, the body could no longer yield up a map of its own death. Rather, it became the site of a less direct hermeneutic project, one that could only be represented to the jury through testimony of those whose professional expertise afforded them privileged access to inner signs.

The role of the viewer and the status of the body as an evidentiary focal point were conceptually altered by the idea of an illegible modern body. By the late nineteenth century, these had become stock methods of characterizing the features of the modern inquest. Criticism of the jury view would almost inevitably gesture to its usefulness in prior epochs marked by flesh wounds, incompetent medical opinion, and local knowledge. The *Lancet* acknowledged the methods of "primitive, but important, inquiries"—involving, for example, "appeal of wounds," which entailed careful measurement of length, breadth, and depth—but urged that "scientific evidence during the intervening period has, of course, developed enormously, and juries listen to the testimony of medical witnesses who inform them with authority as to the cause of death, besides assuring them incidentally that there is a corpse upon which, to use a quaint popular phrase, they are sitting."[48] Accounts of the modern juror's encounter with the corpse reinforced this image of futility. Jurors were trapped in a cruel ritual, enacting their incompetence by "shut[ting] their eyes when they filed into the presence of the dead."[49] The 1909 Parliamentary Select Committee on Coroners heard a particularly well-turned summation of the case against the view: Jurors "go in with their hands up to their noses and their eyes half shut; . . . you might have had a dummy there so far as being able to assert anything as to the cause of death by that view is concerned. It is an absolute farce."[50]

If all jurors subverted the view in this way, the ritual would have remained an objectionable transgression of the divide between the scientific and the public, but it would be inconsequential in practical terms. The problem instead lay with potentially atavistic jurors who, refusing to acknowledge their incompetence before the body, might eschew modern sensibilities and deference to science by actually taking an interested "view." Critics did not have to look far for confirmation of their fears. Medical journals and meetings of professional and political organizations aired a raft of tales featuring jury misinterpretations of the body. A meeting of the Coroners' Society heard its secretary relate the consequences when a jury took their view literally: "On several occasions the view of the body by the Jury, even when a Medical Practitioner has made a post-mortem examination, has been found to be misleading to the Inquiry. Post-mortem staining has been mistaken for bruises; the opening of the head has been mistaken for fractures of the skull, and it has been with the greatest difficulty that the matter has been explained to the Jury."[51] Confusion, misreading, and inaccurate verdicts were the inevitable result of an uninitiated set of visual interpreters of the body, whose tendency to fix on meaningless external signs reflected an archaic order of knowledge.

Disengaging the body from its lay interrogators, finally, held out a promise for addressing the vexed problem of the inquest's physical context. The requirement for a popular view of the body fixed the inquest in a local, public space, thus presenting a logistical impediment to removing inquests to the purpose-built confines of the court and mortuary. With the body physically disengaged and present only in the abstracted form of a medical expert's postmortem description, the inquest might move to a symbolically and operationally purified format. "Owing to the present practice of the jury viewing the body," the LCC observed in its 1895 report on inquest law, "it is a necessity that the inquest shall be held near to where the body lies. But that practice might be unnecessary if a responsible medical officer of the court examined each body and reported the result. The coroner could then sit regularly in one court."[52] Statements like this underscored the pivotal role that the ministrations over the body played in the conduct of inquests and the opportunities for fundamental reform that might follow from economy of display. Conceptual "purification" had an irreducible physical dimension.

III

A "decorporealized" inquest pointed unambiguously in the direction of an expert-based, efficiency-oriented system of death management. This program, however, was consistently mixed with concern for the value of the inquest as a ritual of public participation and expiation. The problems of abstraction lay at the root of these concerns. Having urged the liberation of the body from the profane space of the public house, having proposed a disengagement of the body from a direct encounter with representatives of the public, reformers now had to determine how to ensure an acceptable public transmission of information generated from within the closed space of their purposeful inquest. If the medical view was sufficient, if, in the words of one witness before the 1909 Parliamentary Committee on Coroners, "we have the evidence of surgeons who describe everything to us," how might the system ensure a stable and acceptable representation of the body to the inquiring public?[53] How could it manage the public curiosity and suspicion that served as one of the inquest's central rationales? These questions had profound implications for the future standing of the inquest. Without an enactment of direct public access to the matter of death, the project of reform, intended to produce a disciplined, scientifically purified inquest, might ironically result in a loss of the very dominion of medical expertise over the body that reformers were seeking to cement. One of the consequences of the inquest's hybrid logic and function was the creation of this paradoxical figure—a disappearing body that couldn't quite disappear.

Even the most ardent reformers voiced concerns about the inadequacy of a fully insulated relationship between scientific medicine and the body. The *BMJ*, in its editorial disparaging the jury view on the level of epistemology, strayed off the path of resolute criticism in acknowledging another function of the view: "Perhaps the greatest objection to absolute prohibition of the view is that the public might sometimes suspect the coroner or the doctor, or both, of trying to hush up matters and conceal the truth from them."[54] The physical presence of the body at inquests enacted transparency, demonstrating pure access to all facts of the case under inquiry. While jurors could not make sense of the body, denying them a physical encounter put at risk one of the inquest's exemplary gestures of openness.

If access to the body grounded the inquest at an unimpeachable level of the real, the relationship between the body and its surroundings provided another salutary anchor. The elimination of the view and the construction of courtroom and mortuary complexes, as suggested above, were two aspects of a single reformist trajectory: without a view the jury could be disengaged from a proximate position to the corpse, which in turn would enable the body to be transported from one end of a coroner's often large district to another, efficiently and accurately, in the form of pages in the medical examiner's casebook. By the same token, a properly constituted complex, with mortuary and adjacent court, allowed centralization of the proceedings within a district. Juries would no longer need to be called up from the immediate neighborhood in which the death occurred. This promised exactly the kind of economy, in both monetary and performative terms, that many reformers sought.

But for others, eliminating the view introduced potentially troubling gaps. E. M. Harwood, the Bristol deputy coroner and staunch defender of the prerogatives of his office, argued that convenience, efficiency, and decorum came at the cost of the inquest's essential "local" functions. Coroners were often put in the position of holding inquests on bodies whose deaths they knew to be "natural" even though they were prevented from being certified because of neighborhood suspicion. As Harwood explained, "It is therefore better sometimes to hold an inquest, even if the coroner has himself come to the conclusion that there is no real necessity for it, the death being, in fact, a natural death, if only to put a stop to idle rumours." Because idle rumors were of situated significance and origin, it was imperative that a demonstration of their unfounded nature be performed before local representatives. Furthermore, as the questions generated by rumor might often be matters for local knowledge (concerning the habits and relationships of the deceased, for example), disengaged jurors assembled in an insulated space were unsuitable substitutes. "It is true that it is sometimes very disagreeable and inconvenient to have to sit in small and uncomfortable public-house parlours or taprooms, as we have often to do," Harwood conceded, but these were ultimately the sign and physical guarantees of a properly constituted inquiry.[55] The encounter with the body and the context of its display were material anchors to the inquest, a check on the dangers inherent in disengaging narrative from a stable originary referent.

The introduction of a gap between the body and its story was at the root of other hesitations about dispensing with the view. Some of those who questioned plans for the reconstituted inquest held simply that the view was epistemologically useful, that the coroner and lay jury really could gain a better understanding of a death by examining the mortal remains. This argument tended to be restricted to specific types of bodies—those bearing, in the words of deputy coroner and medicolegal author Richard Henslow Wellington, "marks upon the surface [which] speak for themselves."[56] More common, however, was concern that an inquest separated from a present, physical body would lose its grounding in the real. In testimony before the 1909 Parliamentary Departmental Committee on Coroners, Wellington justified his support of the view by insisting that "the jury would lose a certain interest in their inquiry if they did not view." A sympathetic Sir MacKenzie Chalmers, chairman of the committee, pressed him on this point: "You mean it becomes unreal, a paper inquiry, if there is no view? . . . It is more like reading the thing in a newspaper than an actual inquest?"[57] Wellington agreed. Other witnesses offered evidence pointing to the view's reality effect. George Vere Benson, an East Sussex County coroner, observed that, while the view was seldom important as a matter of evidence, "it gives more impressiveness and more reality to the inquiry."[58] A Home Office memorandum drawn up just before the committee's deliberations had likewise determined that the view was "a solemn ceremony calculated to make jurors take a serious view of their functions." To this detailed document the Home Office's C. E. Troup appended a simple affirmation of the view as an authenticating rite: "It is better for people to see things with their own eyes."[59]

Raising concerns about the medicalization of the inquest could thus easily shade into an embrace of the ideals of public inquiry. In the end it was the public that had to be convinced of the story told about a suspicious death. None of these hesitant voices, it should be noted, spoke in resolute opposition to a reformed inquest. Instead, they expressed doubt about the capacity of a representation of the body emanating from the hermetic domain of medical expertise to serve as an acceptable mediator between the body and its public.

The problems of absence and representation posed by an expert-rendered body intersected with and were reinforced by another set of considerations of an ostensibly more mundane sort. These involved coroners' concerns about authority, about a loss of control of the body not merely as evidence but as jurisdiction. Anxiety over their status both within the English legal and political structure and with the public at large had long been a feature of coroners' corporate identity, and it had been largely the desire to improve their image that had prompted the formation of the Coroners' Society in 1846.[60] Schemes for the promotion of the dignity of their office were a recurrent topic of discussion at the society's annual meetings. In 1905, to take but one example, decades-long discussion about whether to signal the solemnity of their duties by adopting specially designed coroners' robes came to a fractious climax. Proponents of formal regalia argued that the coroners' dealings with the "unenlightened" and "crassly ignorant" members of the community in less than tranquil settings made such distinctiveness in dress a potentially important fulcrum of authority. The Manchester coroner urged that "it might save argument and trouble if the Coroners bore some outward semblance of the dignity and power vested in their office."[61] Others dismissed the idea as a distraction from the fundamentals of dignity-enhancing reform, with one member arguing that any deficiency in dignity was attributable "not to dress but to the buildings in which inquests often must be held."[62] A third faction condemned the proposal as a betrayal of the popular essence of the inquest: "A Coroner's Court," a Kent coroner argued, "had the reputation of being a poor man's court, and Coroners, mixing with the people as they did, were accessible to the poor—an accessibility which would not be enhanced by Coroners sitting aloof in wigs and gowns."[63] The motion was tabled for still further consideration.

The fate of the ancient practice of viewing the body presented coroners with a more troubling set of choices. A decorporealized inquest, on the one hand, might extricate their office from its association with the unsettling materiality of the dead body, the desire for which is evident in coroners' own descriptions of their office. "Let me say at once that a Coroner is not a ghoulish sort of person to be treated with a slight shudder as a grim sort of joke, like an undertaker," Ingelby Oddie rather piteously began his memoirs on a life spent as a coroner in early twentieth-century London. "We are not constantly dabbling in blood. We do not make post-mortem examinations . . . There is no need to shrink when you are introduced to a Coroner. He is quite respectable. His clothes do not smell of blood. His hands are as clean as yours."[64] But if dignity followed from a less direct interaction between the coroner and the body, it was this same connectedness with the body that was the fount of a coroner's authority, the very basis of his office. Once a coroner gave notice that an inquest was to be held, the body fell under his exclusive jurisdiction. He was the only official empowered to commission a postmortem examination, and permission for burial in cases under his jurisdiction was subject to his approval alone. For coroners, bodies were more than evidentiary objects; they were the stuff of a unique civic ministry.

The minutes of the Coroners' Society bear witness to this professional interest in a present corpse. As a matter of broad principle, the members of the society agreed that the view of the body, both by themselves and by the jury, added little to the ostensible purpose of the inquest—attaching a cause to a death.[65] But at their annual meetings they continually rejected proposals either to draft or to support legislation to this effect. Summing up the society's ambivalent position, its secretary, A. Braxton Hicks, acknowledged the operational and conceptual case for elimination. Hicks nevertheless warned his colleagues that "it would be unwise for Coroners, as a body, to do away with the distinctive mark of their jurisdiction." The coroner's control over the physical body was his defining characteristic, Hicks argued, and would protect him from the various projects afoot to reform him out of existence. Other officials, Hicks advised, "would be very averse to have such an unpleasant duty cast upon them for nothing."[66] If, on the other hand, by a reform of inquest procedure the body was considered sufficiently present for the purposes of inquiry in medical testimony alone, interlopers might look to expropriate coroners' duties—and emoluments—for themselves. Here Hicks invoked the central place in the inquest ritual of a fully material corpse, a thing offensive to normal sensibilities. Although in other discussions coroners attempted to neutralize such encodings of the body, in this case they were brandished as a tool of differentiation, as an acknowledged source of pollution that coroners alone could manage in the name of the public welfare.

Coroners were also anxious to maintain the leverage afforded by the view in another battle for authority during the era of mortuary construction. Individual coroners had for decades made it clear that their attempts to remove bodies from private dwellings to newly constructed mortuaries had met with appreciable resistance from relatives and friends of the deceased. In testimony before the 1893 Committee on Death Certification, Hicks observed that "there was a great deal of outcry at one time, because the metropolitan coroners, having got mortuaries, moved all bodies there for inquests; but that has died out; they are brought decently in a hearse to the mortuary and left there, and it looks just as if they were going to the church yard a little sooner than usual."[67] Hicks's account of steady progress in the face of popular prejudice found an echo in the 1909 parliamentary hearings. Manchester's coroner, for one, informed committee members that, when he took office in 1904, "practically no bodies were sent to mortuaries even for postmortem work. When first I introduced the habit of removing them in the case of post-mortems there were riots in the streets sometimes, and when I sent my officers to remove a body, in one or two cases the friends of the deceased rallied and rescued the body and took it away. Very great tact has been required to get matters as far as I have got them." The process of educating the Manchester populace to the benefits of re-movals was not complete, he acknowledged, but progress was being made.[68]

Under these circumstances, coroners employed every means at their disposal to assert their absolute authority over bodies. If it served no other purpose, here at least the view had logistical value—it provided practical leverage for prizing the body out of recalcitrant private spaces. Mortuaries, coroners could (and did) argue, were the only places in which their jury, numbering twelve to twenty-four men, could conveniently, safely, and decently perform their statu-torily mandated ritual.

The minutes of the Coroners' Society reported on a rift between the Southampton coroner and the municipal authorities which per-fectly illustrates this mundane value of the view. In 1896 the town's clerk wrote to the local coroner, William Coxwell, disputing his right to remove bodies from private dwellings and accusing him of "order-ing the removal of bodies from private dwelling houses to the mor-tuary as a matter of convenience to yourself and your juries." Cox-well sent the letter on to the society for advice, accompanied by a note explaining his position. Most of his inquests, he wrote Hicks, were upon Southampton's poorer citizens, "who have no proper accommodation for their dead, whom they would keep, unless re-moved to the mortuary, in the room which is generally the living and sleeping room in which the whole family reside, and this I think perfectly justified, on sanitary grounds, the removal of the body."[69] Coxwell concluded his précis of the dispute with a warning: "The matter will become a serious one if the hands of the Coroner are to be tied as to the removing of bodies." To this assessment the honorary secretary readily assented. It was an established point of common law, Hicks wrote, that while there was no property in dead bodies, they properly remained in the custody of friends and rela-tives pending decent disposal. However, Hicks urged, when the coroner demands inquiry, possession must pass into his hands. And it was up to the coroner to determine the degree of possession required. Coroners' power of removal, Hicks contended, "is unlimited so long as public decency is not outraged." In his concluding sentence, Hicks tied the conflict of possession directly to the mechanics of—and thereby the leverage afforded by—the view: "The houses in which bodies may be lying dead are often times excessively inadequate for the purposes of the view, which is as a rule the chief object of removing a body to a mortuary."[70] In their efforts to assert their right to fully "sit" on a body, coroners might look to the view as a welcome practical ally, irrespective of any evidentiary value that may or may not follow.

Coroners, though of course particularly sensitive to the finer points of corporeal jurisdiction, were not alone in their appreciation of the operational value of the view. The *BMJ*, in the very same editorial that dismissed the view as a farcical ritual devoid of evidentiary value, acknowledged that its complete abolition might lead to a dangerous state of affairs. Careful attention to the question of control was necessary.

> It is of the utmost importance that the body should be absolutely under the coroner's control and legally in his possession, so that he may be able to deal with it as circumstances require, removing it, having it examined by experts, taking photographs, etc., and preventing it from being in any way tampered with or disposed of until he is satisfied that it can yield no further evidence . . . To omit the view by the coroner would be to throw away the one outward and visible sign by which his direct control is manifested.[71]

Preserving the coroner's view of the body allowed the apparatus of inquiry to retain proprietary rights over its most important object. In this assertion of jurisdiction, reformists had to acknowledge their own stake. The body, despite being neatly contained within a theory of exclusive expertise, thus kept slipping out, refusing to dissolve itself into the pages of a postmortem report or the univocal recitation of a medical witness. The jury's view, dismissed as a danger to health, morals, and truth, returned as a sign of the real and a valuable buttress to authority.

TELLING TALES OF THE DEAD

Inquests, Expertise, and the Postmortem Question

Two ways of telling the dead's tales are at work in the following passages, each of which, though sharing certain basic assumptions, ultimately describes a quite distinct narrative regime.

> A dead body tells no tales, except those which it whispers to the quick ear of the scientific expert, by him to be reported in the proper quarter.
> —Douglas Maclagan, 1878[1]

> In numberless cases it is the practitioner in attendance who alone can complete the story. His place as a witness cannot be taken by an independent medical investigator who never saw the deceased in the flesh.
> —*The Lancet*, 1907[2]

In each text the medical witness stands at the center of death inquiry, serving as a supplemental storyteller for an audience unable to draw its own conclusions about an abrupt, unsatisfying death. Each witness reaches into a world of silenced flesh with his own set of interpretive tools, promising to wrest from it one final account. But for the *Lancet*, "flesh" refers to life, to history, and its medical witness approaches the body equipped with a clinical guide. For Maclagan, dead flesh whispers its telling secrets to one trained in its own singular language. In his "scientific expert" diachrony yields to synchrony as the predominant interpretive perspective, with the keys to death's enigma all existing on the present and bounded space of the postmortem table.

In this distinction lies a central dispute about bodies at inquests, a dispute that by the late nineteenth century had widely come to be known as "the postmortem question." Chapter 3 showed how the physical body, as the operational and symbolic center of the inquest system, served as a crucial factor in the campaign to produce a more scientifically satisfying procedure for inquiring into unexplained deaths. By seeking to recast the inquest's spatial and interpretive relation to the body, proponents of inquest reform sought an exclusive encounter between medicolegal expertise and the evidentiary touchstone of the inquest's deliberations. To discipline the context of the physical body was fundamentally to redefine the nature of public inquiry into death.

In this effort there was a level of cohesion—if not at all times a self-confident one—among representatives of the medical community in their insistence that investigations of the dead body at inquests required a more scientific framing. But if broad agreement could be reached concerning the need to supplant the polyvalent spectacles of the pub and the view with the more decorous and productive interrogation of the corpse by a representative of the medical profession, no such cohesion existed when it came to deciding on medicine's representative viewer of the body. From the last quarter of the nineteenth century and continuing well into the following century, the question of who should be charged with operating in the projected disciplined space of the mortuary—that is, who should perform inquest postmortem examinations—opened a breech in the project of medically recasting the English inquest.

I

The roots of the controversy extend back to the terms of the 1836 Medical Witnesses Act. Pushed through Parliament by Wakley in his capacity as Finsbury's newly elected member, this act for the first time set out rules for summoning and remunerating medical evidence at inquests. It authorized the coroner to pay for one medical witness per inquest, at rates of £1.1 for testimony alone and £2.2 if he had issued an order for a postmortem examination. Application could be made to the Home Office for permission to employ additional medical witnesses in exceptional circumstances, but the statutory terms meant that normally the coroner had to choose one witness to embody all (or at least the best) available medical knowledge about the death in question. Failure to comply with a coroner's summons to conduct the required examinations and appear as a witness before the inquest was made a finable offense.[3]

The principle governing the coroner's selection of this witness, according to the terms of the act, was proximity (i.e., a physical connection to the case under question). Ideally, the paid witness should be the doctor who attended the deceased during his or her last illness; failing that, the coroner was instructed to summon the practitioner called in at or immediately after death or, as a last resort, any qualified man practicing in the vicinity. To be sure, a paid witness had to be a legally qualified practitioner—licensed, that is, by any of the numerous organizations that were entitled to grant medical credentials in the period before the educational reforms of 1858. Beyond this stipulation, however, the basis upon which medical witnesses appeared at inquests was no different than that of any other witness. They were selected not for their special skill in reading dead bodies as generic texts, but for their supposed capacity to provide evidence in relation to the specific circumstances of a specific death.[4]

By making obedience to a coroner's summons a statutory obligation based on normal professional involvement with a fatal case, the Medical Witnesses Act placed the ordinary practitioner at the very center of the English system of death inquiry. But, contemporaries worried, was he prepared for the job? For decades common wisdom had held that English medicolegal practice lagged far behind its Continental counterparts. "It has been a just reproach to England," the preeminent medical jurist and toxicologist Alfred Swaine Taylor observed in his first of many treatises on the subject, published just before the passage of the Medical Witnesses Act, "that, although she set the example to other nations of bringing to its present state of perfection, the system of 'trial by jury,' she has allowed them to take precedence in the cultivation of Medical Jurisprudence."[5]

English textbooks traded heavily on the anxiety generated by this system of obligation and expectation, thrusting themselves into what they identified as an educational void and demanding that the subject be taken seriously by the profession at large. The all-too-common assumption that the ordinary practitioner could afford to ignore medicolegal study, they warned, was little more than a dangerous fantasy: "A medical practitioner who thinks himself secure in the most retired corner of the kingdom," Taylor warned, "is liable to find himself suddenly summoned as a witness on a trial, to answer questions which perhaps during a long period of practice he had been led to regard as trifling and unimportant. Under the circumstances it is scarcely possible that he can avoid exposing his deficiencies."[6] In this respect, the 1836 act represented a marketing bonanza for medicolegal instruction, a point not lost on its critics. Indeed, the *London Medical Gazette* condemned Wakley's sponsorship as a crass commercial ploy "to procure purchasers for his journal by holding out his own 'act,' *in terrorem,* over the medical profession. He publishes lectures on Medical Jurisprudence, and then appends as a standing notification, that the enactments of the Medical Witnesses' Bill render it impossible for practitioners to neglect the subject, *'without exposing themselves to the risk of irretrievable ruin and professional degradation.'"*[7]

Concerns about the standards of medical witnessing and the personal and professional humiliations awaiting the ill-prepared practitioner retained their sense of urgency in the second half of the century.[8] But they increasingly came to be voiced as a prelude to a more structural critique, one that invoked skill-based distinctions among potential medicolegal witnesses as the basis for radical inquest reform.[9] This second line of attack capitalized on the heightened institutional profile of pathology and its various subdivisions (including forensic medicine) that the broad specializing trend in late nineteenth-century medicine had helped to create.

By the end of the 1870s, most English commentators would have conceded (some enthusiastically, others reluctantly) the *Lancet*'s claim that pathology in the past twenty years had undergone "an immense revolution both in its methods and results."[10] This was a "revolution" characterized in broad terms by a refinement in both the analytic and the taxonomic levels at which pathologists operated and a relocation of their activities from the dead room to the pathological laboratory equipped for histological, chemical, and even physiological investigations.[11] At least in theory, these developments extended to the preparation required of the modern medical witness. Thus, even skeptics of the "modern" embrace of specialism like *The*

Times, while in other places voicing alarm about a coming "tyranny of *expertise,*" agreed that "the advance of medical science has long forced those who follow it to adopt a somewhat extensive system of division of labour, and the fitness to pronounce authoritatively upon medicolegal questions is not to be obtained except as the result of special training."[12]

In casting medicolegal testimony as an attribute of specially acquired expertise, however, it was the traditional work of conducting postmortem examinations that took center stage. Taylor was in no doubt that amateur autopsies were the critical flaw undermining the accuracy and credibility of medical witnessing and that, by regularly entrusting postmortem examinations to those "not well skilled," coroners jeopardized their inquiries. Yet this was not their fault, Taylor continued: "The error is rather with the system; and the sooner it is abolished, and a more reasonable mode of proceeding substituted, the better."[13] Taylor's call for radical reform was echoed in the last quarter of the century, as forums like the Social Science Association and the British Medical Association outlined proposals aimed at shifting the burden away from the general practitioner.[14] These had won provisional legislative backing when an 1879 parliamentary subcommittee recommended, in the interest of "insuring that post-mortem examinations, which may be requisite for the purposes of an inquest, should be conducted efficiently, the Coroner for each district should nominate one or more competent men . . . by one of whom on all cases the requisite post-mortem examination should be made."[15]

"The postmortem question" had reached this level of notoriety in no small part because of a series of widely publicized inquests that seemed to highlight the inadequacy of existing arrangements. By far the most important of these was the inquest on Harriet Staunton, whose emaciated body was brought before a Kent coroner and his jury convened in the town of Penge in April 1877. Circumstances suggested that the deceased had been a victim of deliberate starvation by her husband and his family, and the local practitioners entrusted with the autopsy pronounced in accordance with this presumption. At the subsequent criminal trial, however, the Staunton defense managed to cast grave doubts on this evidence, arguing that other conditions capable of producing an emaciated appearance had not been considered by the medical investigators. The Stauntons were convicted and sentenced to death, nonetheless. This provoked a flurry of debate; *Lloyd's Weekly Newspaper* noted the "universal controversy" provoked by Penge, while a reporter for the *Penny Illustrated Paper* ventured that "no gathering of Englishmen can (as far as my experience goes) be long together without 'The Penge Mystery' cropping up."[16]

In arguing their case, defenders of the Stauntons embraced the discourse of expertise by drawing attention to the qualitative distinction between generalist and specialist evidence. The Staunton's own petition for pardon, drafted by their counsel and presented to the home secretary, protested

> that a post-mortem examination was held on the body of the said Harriet Staunton by four general practitioners resident at Penge and your Petitioners are advised that such examination was inefficient and the practitioners inexperienced and the results obtained most unsatisfactory owing to the absence of some eminent Pathologist who would probably have searched and discovered the appearances which would have shewn the true cause of death and would thus have demonstrated with greater accuracy the presence of disease.[17]

Widespread expressions of concern at the conclusion of the trial—including a petition signed by over six hundred medical practitioners calling for a thorough reexamination of the scientific evidence—provoked a Home Office review and, ultimately, a commutation of the capital sentences against the defendants.[18] This understandably did little to stem the controversy raging in the lay and medical press, but, the *Lancet* ventured, by demonstrating to all interested parties the need to resolve the postmortem question, the Staunton inquest had served at least one purpose. "No one," it declared at the close of this troubling year for English legal medicine, "is satisfied with the existing state of matters, and public opinion is clamorous for reform."[19]

But what kind of reform, and in what relation to the broader questions over the inquest's future? Postmortem evidence, after all, had long been a staple of plans for medically reconstituting the English system of death inquiry. As noted in chapter 1, postmortem investigation featured prominently in efforts by Wakley and others to place the inquest on a sounder medical footing, with the frequency of postmortems conducted for inquests serving as a gauge of progress in this endeavor. These calls, moreover, had not been without effect. Postmortems were undoubtedly on the rise by the second half of the nineteenth century, a trend welcomed by the great majority of medical commentators on the inquest.[20] But advocacy of more frequent postmortems at inquests was not coextensive with the argument that experts should be entrusted to perform them. To justify a shift in inquest practice toward expert-based testimony, advocates had to constitute postmortems not merely as necessary, but as tasks outside the ordinary practitioner's sphere of competence.

In appealing to an image of postmortem practice as a singular scientific endeavor, advocates of an expert-oriented inquest had available to them strongly argued contemporary pleas in favor of a postmortem regime set free from other medical activities and interests, the most influential being Rudolph Virchow's 1875 classic, *Post-Mortem Examinations*.[21] Postmortems needed to be systematic and analytical, Virchow insisted, arguing further that this could only be achieved by bracketing clinical information on the death under scrutiny. He acknowledged that postmortem examinations were not ends in and of themselves, but for them to serve any useful purpose they needed to be conducted in a more insular context than usually obtained. "The present generation," Virchow observed, "is conversant with pathological anatomy only as a supplement of the clinic. As a rule, the clinical teacher determined while the patient was alive which organ was to be the object of investigation; and the autopsy likewise was usually confined to that organ, or at least dealt with all the others only in a secondary manner."[22] This ran counter to the broader scientific requirement that postmortems be complete, distinct, and conducted in isolation, with clinical clues dictating neither the order nor the extent of the examination.

In his vigorous advocacy of autopsy liberated from clinical teleology, Virchow provided a model for postmortem inquiry as an important and distinctive tool of modern, scientific investigation, one noticed by English commentators. In the wake of the Staunton affair, the *Lancet* urged that autopsies, particularly those "undertaken to clear up a mystery," ought to be conducted "in strict conformity with the most complete methods of scientific investigation."[23] Virchow himself delivered a verdict on the English system in the aftermath of the Penge controversy. Asked by the *London Medical Gazette* "to declare into which scale he throws the weight of his universally recognised authority," Virchow emphatically and not surprisingly endorsed a shift to expert-based witnessing, adding his hope that in so doing he might "in a small degree assist in reconciling opinions across the Channel, and in bringing about the introduction into forensic practice of special knowledge, method, trustworthiness, and a wise reserve."[24]

A parallel discourse of pathological specialism circulating in self-promotional accounts of the subject reinforced this emphasis on a systematic and independent set of postmortem protocols. Here the pathologist's special authority to declare on matters of death derived from a rigorously enforced delineation of his sphere of competence, through which he was in effect identified with the material solidity of his investigative object. The 1846 charter establishing the Pathological Society of London, for instance, expressly confined the society's sphere of activity to the exhibition of specimens from the postmortem room. Meetings were to simulate as closely as possible the unmediated experience of the autopsy, an objective underscored by the stipulation that "in the remarks made in reference to specimens exhibited, all discussion on abstract topics shall be as far as possible avoided."[25] This apotheosizing of things-in-themselves was the pathologist's stock in trade, a rigorous standard that, in restraining impulses to abstract theorizing, guaranteed the direct transmissibility of his findings. Half a century later, the society's newly installed president professed continuing support for this founding principle. Limiting discussion to things actually present before the society, Joseph Payne observed in his 1898 inaugural lecture, functioned "like the rules of admission of evidence in courts of law, which, though sometimes excluding materials useful for arriving at a verdict, yet ensure that the evidence which is admitted shall be unchallengeable. There is a vast difference between the statements of any observer, however competent and veracious, about an object, and the view, handling, or examination of the object itself. The subject matter of science," Payne gravely observed in closing, "is not statements about things, but the things themselves."[26]

II

The skilled witness specially trained in methods of postmortem examination served in the debate about the nature and function of the coroner's inquest as the personification of objective, removed, insulated inquiry into the cause of death. In his role as a harbinger of a reconstituted inquest, the expert pathologist's defining characteristics were developed in relation to and reinforced a specific set of criticisms about the general practitioner–centered inquest. The special pathologist stood in stark opposition to the undisciplined, porous inquest, anchoring an alternative vision of inquiry based on disconnectedness, demonstration, and purity.

As a structural feature of their work, general practitioners were regarded as ill-equipped to adhere to the basic tenets of scientific postmortem practice. In their critics' eyes their examinations violated the proper boundaries of the postmortem room, admitting the pathologized body as an individuated case and an object of sentiment. When asked to perform postmortems for inquests, for a start, general practitioners all too often shunned "full" investigation, sacrificing science to sentiment by yielding to "entreaties of relatives and friends not to extend the examination further than was absolutely necessary."[27] The practitioner's sacrifice of postmortem protocol in the face of such extrascientific pressures was fully understandable, as any doctor who developed a reputation for excessive zeal in probing bodies might soon find himself without a client list. But this merely confirmed the need to entrust examinations to specialists whose sole client was the public and whose imperviousness to local and private sentiments could be translated into a scientific passion for accuracy.

The connectedness of the general practitioner also undermined rigorous analytical method through an interposition of prior clinical knowledge. Enthusiasts of postmortem protocol, of course, did not maintain that the examiner ought to be wholly ignorant of a case's antemortem particulars, insisting instead that such information ought to be kept firmly in the background while the autopsy was being performed. Such a disciplined analytical approach was singularly lacking in the ordinary practitioner's examination. The practitioner who had attended the deceased during the course of his or her final illness, his critics charged, tended to overdetermine his reading of postmortem signs by referring them to a preexistent code, his own clinical diagnosis. As the medical coroner for Middlesex, Edwin Lankester, observed in his annual report for 1868: "I have often to complain of the imperfection of *post mortem* examinations. The medical man, previous to the examination, having a theory as to what the cause of death may be, satisfies himself with examining only those viscera which he suspects have caused death. If he finds sufficient lesion in any of the viscera, he gives his opinion as to the cause of death."[28] Critiques of the circularity of the postmortem hermeneutic when left in the hands of the attending doctor thus identified the lack of objective insularity from the context of death as the basic flaw of the practitioner-based autopsy. A deferral of interpretive closure, particularly when faced with the need to satisfy demands for a clear cause of death in the context of a public inquiry, could not be expected of an ordinary medical man.[29]

In making their case, critics of the amateur autopsy also had available to them the powerful discourse of public health. The dangers of postmortem infection due to unregulated interaction between medical practitioners and the dead had been brought to the attention of the profession by midcentury, most notably through Semmelweis's etiological study of puerperal fever in hospital maternity wards. Interdicting access to the corpses' interior to all except its specialized interrogator could be construed, on these grounds, as a sanitary imperative. The specter of pollution emanating from the want of proper confinement of the postmortem examination could be evoked to disturbing effect, as when a provincial general practitioner wrote to the *British Medical Journal* (*BMJ*) observing "how very uncomfortable a man must feel to receive an urgent summons from such an examination, with his hands steeped in necrotic material, and his whole person polluted with the miasma, to the bedside of perhaps a lingering primipara, and what finessing there must often be in order to gain as much time as possible that disinfection, imperfect at the best, may be duly carried out."[30]

Images such as these unequivocally cast postmortem examinations as encounters with polluted matter, requiring special agents whose circumscribed social and professional position localized their contaminating effects. This extended the contemporary discussion on the authorized "viewers" of the body analyzed in chapter 3: here basic medical credentials were not in themselves sufficient to ensure a pure and purposeful encounter between the dead and the living. The proper border was more restrictive, admitting in effect only a priesthood of pathologists, distinctive as much in their ability to mediate between bodies and the community as in their special brand of knowledge. Standing above the thoroughly compromised general practitioner was the idealized figure of the expert pathologist. He would confront the dead body as the object of unmediated contemplation. He would have neither prior knowledge of the corpse nor any interest in the findings other than the demonstration of his skill, which mapped directly onto the interests of truthful inquiry. His livelihood would be derived from a state salary and not from the continued patronage of any constituency. His professional activity would be contained within the purposeful space of the mortuary, and his contact with the external world would be limited by both the logic of his practice and its polluting potential.

But, as in the case of inquest reform writ large, this purely medicalized solution even here was not the dominant position adopted, and this for two basic reasons. First, general practitioners had an acknowledged place in the public performance of inquests, one that, as will be discussed toward the end of this chapter, could not be dismissed as a straightforward anachronism. The second impediment to an unambiguous embrace of expertise was more subtle: from within the very discourse upon which the claims of distinctiveness were based—in the pathological textbooks and postmortem manuals—a clear "public" sensibility sat alongside and destabilized claims to a resolute hermeticism. In representing expertise, in other words, English medicolegal texts reproduced the same hybrid space occupied by the inquest and its peculiarly English way of investigating death.

This complexity is evident in discussions about how the pathologist's direct encounter with the body could be replicated before a public tribunal of inquiry, how his concrete knowledge could be represented without losing its fundamental attribute of transparency. Though they considered this a crucial medicolegal question, commentators like Alfred Swaine Taylor tended to give would-be experts ambiguous advice. On the one hand, Taylor criticized witnesses for using terms of everyday description in seeking to direct the public mind to the salient features their investigation: "In the drawing up of an ordinary postmortem examination, the lining membrane of the stomach is described as being 'intensely' inflamed; or some part is 'considerably infected,' or a cavity is 'enormously' distended. Expressions thus loosely employed convey to the legal mind a widely-different meaning from that intended by the reporter." But at the same time Taylor cautioned against the self-subverting effects of a literal enactment of expertise, against evidence "overloaded with technical and therefore unintelligible terms." If a medical witness used terms like "'parietes of the abdomen,' 'epigastrium,' 'hypertrophy of the liver,' when it would require no more trouble to put what he means in plain English," Taylor warned, "he must be prepared to have his meaning perverted or wholly misunderstood." This practice, Taylor concluded, "is not science but pedantry; if such language is employed by a witness with a view of impressing the court with some idea of his learning, it wholly fails of its effect."[31]

So foreboding were the pitfalls besetting the public display of expert knowledge and so great the cost of succumbing to "perversion" of meaning that more than one medicolegal writer described the task as an "art" that necessarily pushed the limits of truth and accuracy. The renowned Scottish forensic pathologist, Henry Duncan Littlejohn, combined a familiar plea for descriptive economy with the following observation concerning the sculpting of expert narrative:

> An intelligent reporter, indeed, shows his knowledge of his art as much by what he withholds as by what he gives in detailed description. An account of a post-mortem examination may be stated at tedious length, and to an unpracticed eye appear from its bulk to be exhaustive and

complete. Yet it may be overburdened with details of not the slightest consequence in determining the cause of death or the nature of the case, and the very appearances which alone could settle the question, and which were undoubtedly present, remain unnoticed. The reporter feels himself at sea, and, in his desire to mention everything, allows points of the greatest moment to escape his notice.[32]

The art of analytical and descriptive selection, for Littlejohn, was critical both for the coherence of the would-be expert's investigation and for its public efficacy. Stanley B. Atkinson's *Golden Rules of Medical Evidence* went further still, emphasizing the fictions of evidence to such an extent that they implicitly became the primary structure upon which evidence hung. "Cultivate the power of expression and of repression," Atkinson urged his readers, exhorting them to take advantage of the rhetoric of plain speech to create an illusion of clarity: "The jury will think they understand 'alcoholic disease of the . . . ', 'bad disorder', 'black and blue', 'blood clot', 'blood poisoning'."[33]

The point being made here is not so much that medicolegal writers broached the subject of display, a fact that should come as no great surprise given that medical jurisprudence was irreducibly tied to its public applications. It is, instead, that techniques of display were theorized as coextensive with expertise itself and, further, that the hermetic ideal of postmortem specialism was destabilized—though not necessarily negated—by the exigencies of medicolegal performance. Indeed, medicolegal handbooks from the early nineteenth through the early twentieth centuries evince such a preoccupation with the relationship between scientific and public knowledge that they take on the characteristics of courtesy manuals as much as those of technical tomes. In one of the earliest of these English texts, John Gordon Smith stressed the importance of evidentiary "decorum," and later works continued to lay stress on manner.[34] W. G. A. Robertson recommended that "the medical witness should school himself to an equable mind, and no matter how irritating the questions or the manner of putting them may be, he should maintain a dignified and gentlemanly manner. By doing so his evidence will be much more convincing." If he followed Stanley Atkinson's "golden rules," a witness would be "courteous, dignified, and withal good humoured," having been advised that his *"personal* disposition will count for more with a jury than [his] professional position."[35]

More surprising, however, is the recurrent evidence throughout these texts of fissures within their depictions of integral expertise itself. Indeed, at the core site of expertise—at the postmortem table itself—concerns about the permeable boundaries between the scientific and the public persisted. The postmortem room was ideally a closed space, insulated from all but the purely scientific. Henry Harvey Littlejohn (Henry Duncan Littlejohn's son and a leading figure in Scottish medical jurisprudence in his own right) included in his list of procedures and methods for conducting medicolegal examinations an exhortation to defend the purity of the investigative site: "No one should be allowed to be present at the examination out of mere curiosity or unless specially authorized."[36] Atkinson also included boundary patrol among his rules: "Exclude lawyers and curious laymen, but invite another medical man, especially if your own previous actions may be in question."[37]

In practice, of course, the border between lay curiosity and scientific interrogation was difficult to enforce. But even in theory the ideal of physical hermeticism proved elusive because oscillation between the discourses of exclusivity and permeation was written into the very fabric of specialist texts. While presenting the autopsy as a skilled practice, medicolegal textbooks were nonetheless forced to acknowledge that autopsies took place in everyday settings and under conditions less than exclusive. This point can be quickly illustrated by reference to discussions of postmortem instruments.

If specialized tools can function as a mark of distinctiveness, the tool kit of the pathologist specified in postmortem manuals—a mixture of purpose-built instruments and tools from everyday life— tended to undermine his capacity to claim a special status. That this was true well into the twentieth century is indicated by the items listed among J. M. Ross's required implements for the pathologist of the 1930s: "Short knife with broad blade (postmortem knife); Long knife with 7-inch blade (brain knife)"; "Ham knife"; "Scalpel and dissecting forceps"; "Small butcher's saw"; "Chisel and mallet."[38] Even when the recommended tools were exclusively medical in nature, as in the list enumerated by G. Sims Woodhead, professor of pathology at the University of Cambridge, concessions were allowed in the name of expediency. Woodhead acknowledged that "post-mortem examinations have often to be made without many of the above instruments (and the lack of them should never be put forward as a reason for not making such examination), but these, or decent substitutes for them, should be obtained if possible."[39] Woodhead then proceeded to specify a series of such decent substitutions. Instead of a postmortem slab or stone or marble, he advised, use a firm kitchen table or, failing that, "the coffin lid, or a door removed from its hinges and supported on a couple of chairs." Sanitary considerations dictated that the hands of the operator should be washed with turpentine and then "anointed" with carbolic oil, but if this was not at hand, purification could be performed "with olive oil or lard."[40]

While expertise might not in practice inhere in instrumental exclusivity, it was still possible to argue that skill transformed everyday objects into highly specialized tools. H. D. Littlejohn, in discussing the proper procedures for examining the head, for instance, acknowledged that "no instrument has ever been devised which effects the separation of the skullcap better than the ordinary saw, but," he observed, "to ensure dexterity in this operation, certain precautions must be attended to."[41] Littlejohn then outlined in some detail the type of incision required in such an examination, the level of force to be exerted, the direction of the blade, and so on. In so doing, he took as his marker of expertise the capacity of the pathologist to maintain the integrity of death as displayed on and in the body even as he cut, sawed, and hammered. Yet even the most skilled operator had to be aware of the dangers of using such nonspecialist tools, not so much because they might compromise his investigations but because they exposed his actions to alternative interpretations:

> No practice is to be more reprobated than the employment of a hammer or wooden mallet to crack the internal table and separate the skullcap. Should the cause of death prove to be injury to the brain and fracture of

the skull, this violent procedure must always leave us in doubt as to the exact causation of a fissure in the cranial bones; and we are sure that were the counsel for the defence, in a case of murder by injury to the skull, to produce a mallet in open court, and ask the medical witness to describe to the jury the procedure adopted in examining the cranium and detecting the fissure that led to the fatal haemorrhage, considerable doubt would not fail to be raised as to the exact cause of death and the responsibility of the prisoner.[42]

Even if the pathologist could use implements of everyday life while maintaining the scientific integrity of his proceedings, Littlejohn warned, he was subject to the public's refusal to credit his power to transubstantiate the mixed tools of his trade.

The argument for inquest reform centered on expert postmortems, then, was fraught with conceptual and practical ambiguity. Claims for specialist interrogations of the dead body associated method-ological, professional, and personal insularity with the necessary conditions for the production of accurate causal accounts of death, yet they sat within a wider medical and medicolegal discourse that consistently drew attention to the limits of expert hermeticism. Even texts that regarded the specialist autopsy as a salutary development in the evolution of the ancient inquest implicitly or explicitly repro-duced the complex relationship between the scientific and public constructions of the inquest in their discussions, thereby destabiliz-ing any unequivocal advocacy of the expert position that might have been intended.

III

The case for an expert-based postmortem had explicit—if at times less than perfectly resolute—backing as a step in the reform of the inquest in particular and of English legal medicine more generally. The practitioner-centered postmortem, by contrast, was not com-monly discussed in such programmatic terms, lacking as it did an articulate and proactive medicolegal constituency. And yet the case for the generalist autopsy was forcefully put, not as a matter of ab-stract principle but in response to particular, practical challenges to its legitimacy. Arguments for the ordinary practitioner's rightful place at the center of the modern inquest were thus generated out of specific conflictual contexts, none more significant in its effects than that sparked by London coroner John Troutbeck in the opening decade of the twentieth century. Troutbeck's seemingly immoderate embrace of expertise provoked a vigorous backlash, one that combined attacks on his specific actions, his choice of pathologist, and his preference for postmortem over clinical evidence, for example, with a broader critique of an inquest driven by the demands of self-regarding scientific specialization.

Troutbeck's radicalizing influence should first be placed in its administrative context. Oversight of the late nineteenth-century inquest had, under the 1888 Local Government Act, shifted from county quarter sessions and municipal corporations to newly formed county and borough councils. Of these, it was the progressivist London County Council (LCC), which stood at the time as the world's largest municipal authority, that set out the most innovative and conspicuous proposals for inquest reform. In the first two decades of its existence, the LCC's Public Control Committee (PCC), under the chairmanship of the noted medicolegal reformer W. J. Collins, attempted to reconstitute every significant aspect of metropolitan inquest practice.[43]

In 1895 the PCC turned its attention to the postmortem question. Agreeing with earlier critics that the 1836 Medical Witnesses Act erroneously assumed that in the autopsy room "all medical men are equally competent," its report on inquests proposed to attach specialists possessing the "highest qualifications" in postmortem procedure to coroners' courts. These experts would be expected to perform any medical investigation deemed necessary and to report their findings to the coroner. The coroner could then, according to this scheme, use the expert findings in deciding whether an inquest was necessary.[44]

The PCC quickly followed up its proposal by assembling a list of London pathologists willing to perform coroners' postmortems at the two-guinea fee stipulated by the 1836 act (and still in force).[45] But the PCC's vision for postmortem reform entailed more than administrative action of this kind. Taken to its fullest application, the plan required a transformation of the procedures through which coroners commissioned postmortems, a shift in the very basis on which scientific investigations of dead bodies in the public interest were legitimated. By proposing to grant coroners the authority to commission a postmortem examination as a feature of their preliminary inquiries, the PCC was in effect challenging a basic tenet of the popular inquest—that any death worthy of inquiry was by definition a death to be inquired into publicly by a jury. A system in which an expert pathologist was empowered to conduct a postmortem on a body as a preliminary investigation in effect positioned the examination as preemptive evidence. The postmortem would become the prior term of inquiry, with the body given the chance to tell its story through a private and scientific appraisal of perceptible organic dysfunction.[46]

These proposals, despite their far-reaching implications for the role of ordinary medical evidence at inquests, were initially met with cautious support in the medical press. The promised results of the reform, after all, were attractive in theory to even those inquest commentators who opposed the principle of recasting inquest testimony within the framework of expertise. The editors of the *BMJ*, for instance, counted among its beneficial results a "considerable diminution in the number of inquests," an enhanced "dignity of the coroner's court," and, finally, the guarantee that "*post-mortem* examinations would always be made by competent pathologists—a reform devoutly wished for by many members of the profession, despite the pecuniary loss thereby entailed."[47]

Within a decade, however, John Troutbeck had almost single-handedly turned the *BMJ* and other representatives of the broad interests of the medical profession into determined opponents of the expert-based inquest. Troutbeck had been the legally trained Westminster coroner for fourteen years when the PCC appointed him in 1902 to succeed the late A. Braxton Hicks as coroner for the far larger and more lucrative South West London district. The PCC selected Troutbeck in part because of his willingness to support its vision of a new system centered on expert pathologists.[48] In Troutbeck the PCC got all it bargained for and a good deal more; his uncompromising pursuit of specialist postmortems almost immediately polarized the terms of debate.

The first rumblings of discontent were sounded in the pages of the *Lancet*. A brief notice in the 29 November 1902 issue complained that the new South Western district coroner had passed over a local practitioner in favor of an outside expert, one Dr. Ludwig Freyberger. This initial notice rather unselfconsciously framed the question in terms of professional solidarity and medical proprietary rights at inquests. It recommended the case to the attention of the Coroners' Society, lamenting the "distasteful position" in which Dr. Freyberger was placed by the coroner, "the position of taking fees for work which a professional brother was in a fitter position to discharge than he was himself."[49]

Ostensible solicitude for the embarrassed professional sensibilities of Freyberger soon gave way to a characterization of the Freyberger-Troutbeck connection as a sinister assault on the medical profession, the inquest, and English liberties writ large. Troutbeck was denounced as a radical centralizing reformer, willing to supplant the native tradition in favor of a superimposed Continental rationalism. The *Lancet* criticized his authoritarian streak, which was represented by his willingness to transform the law of inquests "without the aid of Parliament."[50] George Bateman of the Medical Defense Union cast him as a pawn in the LCC's broader aim to reform through administrative fiat and warned coroners, "if they desire to retain the dignity of their ancient and honourable office, [to] stand by the Coroners Act, and refuse to make terms with any appointing authority."[51]

Troutbeck's choice of the Austrian-born Freyberger fed the charges of a Continental encroachment upon the English way of managing death. At times the medical press called attention to Freyberger's foreignness by an affectation of indifference: thus, the *Medical Free Lance* introduced its readers to the unfolding controversy by asserting that it had "no concern with Ludwig Freyberger's nationality."[52] Others were less demure in sounding the call to patriotic opposition. In a speech delivered before the Battersea Municipal Alliance, Dr. McManus, a practitioner in Troutbeck's district, challenged the wisdom of expertise construed as independence from local influence by linking it to Freyberger's native proclivities: "Mr. Troutbeck had tried to find a gentleman in the district to conduct post-mortem exami-

nations who had no local connection. Failing that, he was more fortunate in foreign parts, and he had succeeded in obtaining the services of a gentleman from a country whose ideas, politically and
socially, were NOT THOSE OF ENGLISH PEOPLE." This unholy
alliance had supplanted the open and cooperative relationship between coroner and local doctors with one marked by rancor and
shrouded in secrecy, as local practitioners now received their first
notice of inquests in press reports "stating that a Dr. Freyberger had
made a post-mortem examination, and had attributed the death to
some particular disease or another."[53] The connection between Freyberger's native authoritarianism and Troutbeck's "closed" proceedings were, according to McManus, of a piece. Privileging post-
mortem evidence over the doctor's clinical testimony produced an
inquest in which verdicts were delivered from the enclosed domain
of expertise rather than recounted from a local site in a language
accessible and useful to the public.

Charges of creeping continentalism sat alongside a discussion of
the restricted significance of pathological expertise. Again, criticism
of Troutbeck's innovations was directed both at his specific choice
of Freyberger and more broadly at the purported value of expert
postmortems. Freyberger's credentials as an expert were in themselves disputed. At a May 1903 meeting between Troutbeck's leading opponents and the lord chancellor, for instance, a delegate from
the South West London Medical Society announced that "Dr. Freyberger was not known to [my] Society as a pathologist, and he did
not hold the office of pathologist to any large London hospital, nor
had he been appointed in any public way."[54] References to Troutbeck's "so-called expert" became a stock characterization of Freyberger in the years that followed. A *Lancet* correspondent went so
far as to insist that Freyberger's claims to expertise were negated by
the very fact of his regular participation at inquests. Expertise was
a rare attainment, the writer explained, "and it should be no more
possible to offer them a fee of two guineas for their services than it
would be to offer, say, Sir Edward Clarke the same sum as a fee with
brief."[55]

Freyberger's quest for recognition met with several public setbacks in his decade-long association with Troutbeck's court, none more striking than the charge laid against him by a Wandsworth general practitioner at an inquest in February 1903. Dr. Kernahan, protesting to Troutbeck at being denied what he regarded as his right to "get [the] post-mortem" on the body of a former patient, challenged Troutbeck's references to Freyberger's expertise by dropping the following bombshell: "I should like to make one statement which I think the jury ought to know. Dr. Freyberger, in conducting this special post-mortem, NEVER TOOK OFF HIS OVERCOAT, or touched a single organ or limb of the body." Called upon to explain, Freyberger assured the court that this seemingly anomalous practice was, in fact, quite normal, one that was "adopted in a great many hospitals in making post-mortem examinations." His own physical infirmities had forced him to follow this course in some thousands of cases, he revealed: "I suffer from sore fingers, and I employ the post-mortem porter, who is paid by the Borough Council, to cut open the bodies. I direct him what to do and how to do it. He takes the organ, cuts it open, and also weighs parts of the body. Everything is done that is necessary. There is absolutely nothing that could escape my notice. It is all carried out under my personal supervision."[56]

Kernahan roundly attacked Freyberger's revision of the assumed foundations of postmortem expertise, protesting that never in his experience had he seen a pathologist, in special cases, allow an assistant to open the bodies. "It was absolutely necessary," he maintained, "to take the organs one after another and personally examine them." The response of the medical press to this exposure of mediated expertise, however, proved more muted, largely because the participation of mortuary porters in postmortem examinations was widely acknowledged, if rarely discussed publicly.[57] The *Lancet*, for example, could only muster a half-hearted condemnation of Freyberger in its gloss on the skirmish: "We see no harm in the porter occasionally doing the manual work, or portions of the manual work, of the investigation under Dr. Freyberger's, or preferably under the medical adviser's, eye, but think it unfortunate that Dr. Freyberger should not be in a physical condition to touch the corpse. An attendant can not take the place of a skilled pathologist: *tactus eru-*

ditus is required to detect the friability of an organ or the density of an adhesion, etc., and no coroner ought to look for such information from an attendant."[58] Still, the episode could hardly have advanced Freyberger's cause.

Troutbeck defended his pathologist against specific interventions like Kernahan's, but also sought a broader framework of justification for his routine reliance on a particular expert. Experts were not born to a distinct, elevated status, he explained, and could not be plucked out of the air solely for cases of special difficulty. They were, instead, forged in the drudgery of routine practice. In his emphasis on the practical foundations of specialism, Troutbeck aligned himself, as his critics charged, with those who cast their eyes longingly across the Channel. The stunted development of English legal medicine as compared with Continental practice, Troutbeck maintained, stemmed largely from the uncoordinated manner in which English medicolegal postmortems were commissioned. "All other civilized countries have official Pathologists for enquiries of this nature," Troutbeck wrote in a 1905 letter to the secretary of the Coroners' Society. "England alone lags behind, with the result that in Medicolegal work we have much to learn from our more fortunate neighbours. Medico-legal Pathology is not learnt in a day. There must be both a school for teaching it and an attractive career for men who take it up. Neither of these conditions exist at present."[59] The coroner's court, he advised the Medico-Legal Society the following year, "is an ideal ground on which to build such a school," adding with a flourish sure to raise nativist hackles, "I, for one, see no reason why we should not in time possess our English Professor Brouardel."[60]

Despite his ostensibly impeccable medicalizing credentials, Troutbeck insisted that he was acting in the interest of English liberties as enshrined in inquest law. He was a staunch defender of inquest jurors and described the inquest as a "people's court" to which ordinary citizens could turn when seeking redress for fatal abuses visited upon them by powerful interests.[61] Employing an expert pathologist was wholly compatible with this orientation, Troutbeck insisted, an argument he most strikingly illustrated by references to the contentious subject of overlaying. Overlaying was a verdict regularly delivered at inquests into cases of infant death. This verdict linked a class- and gender-based pathology of drunkenness with the scourge of infanticide by concluding that inebriated (and typically impoverished) mothers had rolled over onto their infants while sleeping in the same bed and suffocated them.[62] By the turn of the century, under pressure from such organizations as the Society for the Prevention of Cruelty to Children, Parliament had under its consideration amendments to the Children's Act which, among other provisions, would place the onus on mothers to prove their sobriety in cases of apparent overlaying.

Troutbeck used his position as coroner to call into question the very existence of overlaying, and in this cause Freyberger played a critical role. In an address before the Medico-Legal Society in 1905, Troutbeck made clear both his opposition to the verdict of overlaying and his views on the source of such misinformed opinion:

> Since I have ceased to employ ordinary practitioners to make the post-mortem examinations in such cases, I find that the number of so-called "overlaying " cases has steadily diminished. The huge majority of such cases are due to natural diseases. It is intolerable that innocent parents should have the stigma put on them of having suffocated their child. Indiscriminate abuse of the labouring classes and charges (often unfounded) of drinking habits have led to an impression being formed that infanticide by means of overlaying is a common thing. The ignorant post-mortem examination seems to confirm the impression. We want more light and less prejudice.[63]

More light and less prejudice were for Troutbeck secured through a well-constructed system of expertise operating within the framework of a "people's court." In this sense Troutbeck—despite his credentials as a municipally appointed lawyer who cut more of a technocratic than a political figure—can be seen as successor to Wakley. Both assumed a compatibility between the public and the scientific, and both looked to the modern medical knowledge of their day as the surest way of preserving and extending the inquest's participatory functions. Thus, though a turn-of-the-century coroner dogmatically committed to a regime of specialist evidence might at first glance seem to represent a straightforward triumph of the modern and the scientific over the traditional and the popular, Troutbeck stands as a variant on a persistent theme, distinct in its historically derived components but embodying the continued power of the interaction between the two distinct founding ideologies of the modern inquest.

IV

The storm created by Troutbeck's actions provided the catalyst for articulating a determined and sustained critique of specialist-centered medical evidence, one that extended beyond the immediate controversy to challenge the broad epistemological and operational claims put forward by supporters of expert pathologists. By, in effect, reversing the terms of opposition deployed within the discourse on expertise, critics sought to call into question the very foundations of forensic pathology as a complete and reliable source of information at inquests. In place of the pathologist as the embodiment of skilled objectivity, his opponents constructed an alternative scenario of overly detached specialism, one that neither guaranteed an accurate determination of the cause of death nor served the peculiar needs of inquest investigation. Instead, it was the very situatedness of the general practitioner, his involvement with the broad features of the story of any given death, which made him the appropriate linchpin of medical testimony at inquests.

Although the pathologist's specialist knowledge of death figured as an important asset in constructing the profile of expertise, defenders of the practitioner's participation in all phases of medico-legal investigation argued the radical limitations of such a disciplinary focus. In making this argument they used whatever material they had to hand, including, paradoxically, the very framework of specialization that they were in principle resisting. In 1909 the experimental pathologist Harrington Sainsbury wrote "A Plea for a More Living Pathology," criticizing the stranglehold that a generalist, dead-end morbid anatomy was exerting over more dynamic research vistas. The *BMJ* turned his plea into an opportunity to pursue its campaign against Troutbeck. According to Sainsbury, traditional pathology lacked the vision and curiosity to go beyond the emblematics of death causation—an "offending organ"—to "discover how the man lived in spite of the obstacles which the specimen exhibits."[64] This, for the *BMJ*, was sufficient pretext for drawing attention to specialist-based inquests: "In many cases, indeed, we are so incurious or our curiosity is so essentially of the type of What, that when we speak of pathologists pure and simple, our ascriptions of the cause of death—though they may satisfy Mr. Troutbeck as long as they are furnished by an 'expert'—are often no more really elucidatory than would be a statement that the patient died from want of more breath."[65] The purity of pathological testimony, the *BMJ* concluded from Sainsbury's article, was a highly stylized but ultimately self-referential mode of description, and the special pathologist—rather than securing a more accurate causal account of death—in fact threatened to return the inquest to the days of formulaic verdicts like "Death by the Visitation of God."

The limited nature of his physical object of investigation further compromised the pathologist's competence at inquests. His unmediated relationship to the dead body guaranteed his objectivity, according to specialist accounts. For his critics, however, this limited relationship meant only that he tended to invest the body with an overdetermined significance, seeking to ground all causal explanation in the physical traces he found upon and in it. This tendency was especially evident in inquest cases, F. J. Smith suggested in his 1906 address to the Medico-Legal Society. Confronted with press and public expectations of a definitive account of death, medical witnesses all too often responded by stretching the already elastic interpretive potential of evidence as to the true cause. The expert pathologist, a "stranger" to the history of the case, "magnifies natural appearances into morbid ones, and makes a statement accordingly—a statement which nobody cares to or perhaps can controvert. How such cases occur is, after all, not very difficult to answer," Smith continued, "for if one takes a general survey of all the possible causes of death, one is immediately struck by the very few absolutely tangible and definitely certain causes of death of which the naked eye makes us at once aware."[66] Pressured by the expectations surrounding an inquest and confronted with an object of inquiry pliant to a range of sanctioned conclusions, pathologists, not merely through ignorance or carelessness but as much by the logic and limits of their enterprise, were tempted to give their findings a false appearance of clarity and surety.[67]

The general practitioner who presided over individual deaths and who was entrusted to perform the postmortem was in a position to avoid this interpretive fallacy. Knowing the clinical history, he could ignore the false signifiers of death and effectively order the multivalent signs present in the corpse into a truthful account of death. This was the substance of a leading article in the *Lancet* rebutting H. H. Littlejohn's uncompromising attack on nonspecialist necropsies. "In the great majority of cases which come under the notice of the coroner," the *Lancet* insisted, "the medical practitioner who attended the deceased during life is the one who should perform the necropsy. His knowledge of the previous history of the case will lead him especially to investigate certain points which an expert not possessing a full acquaintance with the clinical features observed during life might pass over."[68] Far from imposing a clinical teleology that distorted a positivist reading of postmortem signs, a practitioner's familiarity with the case was crucial in bringing order to the chaos of morbid appearances presented by the corpse.

The defenders of the general practitioner's competence to perform coroners' postmortems did not leave the expert pathologist completely out of the picture, however. The place of the star medicolegal witness in delicate criminal cases—notably that of the toxicologist in instances of suspected poisoning—had long been acknowledged.[69] Even beyond cases of "special difficulty," many were willing to accord the expert pathologist a more regular place at inquests, but one strictly subordinated to the deceased's medical attendant. "A pathologist, however eminent he may be," a correspondent to the *Lancet* insisted, "should never determine the natural cause of death before a jury if there has been a medical attendant in the case. The latter should determine, being assisted by another pathologist when desirable, whose only duty it should be to present facts and figures regarding what he finds."[70] By defining the pathologist's role as primarily manual and correlative—properly limited to "facts and figures"— his critics turned the image of positivist objectivism back on itself.

Pathologists were indeed capable of weighing and measuring, but not of interpreting the significance of the data produced by their inquiries.

This desire to establish a restricted remit for the pathologist points to a contemporary tension in the image with which he and his work were associated. On the one hand, the established medical community voiced fears that early twentieth-century pathology was threatening to overwhelm other frameworks for determining medical truth. The *Lancet*'s 1900 editorial on "The Clinical Value of Pathological Facts" commended a recent criticism of the growing tendency "to apotheosise the morbid anatomist who, standing high in his suit of sable, looks upon the clinical surgeon as if his chief duty were to supply him with material."[71] At the root of the clinicians' suspicion was a rejection of the very ideals of disinterestedness and isolation upon which proponents of expert pathology staked their claim to objectivity. For them, such contemporary developments as the proliferation of pathology laboratories were emblematic of the dangerous trend toward insularity in modern medical practice.[72] At the same time as warnings were being sounded against the image of the pathologist as a modern diviner of medical truth, however, pathologists and their advocates bemoaned the "low esteem" with which they were viewed by the majority of their medical colleagues. Pathologists described themselves as medical "Cinderellas," the "sweated labourers" of the profession who were victims of the British tendency to "disfavour of 'scientific' methods as compared with 'practical'."[73] The pathologist's status was contradictory; he was cast at once as an ascendant oracle and an exploited technical laborer.

The epistemological rebuttal to the arguments put forth by the advocates of specialist pathologists thus interlaced a series of themes: warnings against the positivist fallacy of pathological evidence, emphasis on the significance of diachronic knowledge within the postmortem room, and a relegation of the pathologist to the position of a noninterpretive collator of the corpse's physical state, most importantly. The general practitioner was from this perspective not only competent but preferable as the representative of the medical community entrusted with the task of postmortem investigation.

The general practitioner's multidimensional role in a case under his supervision, moreover, helped him beyond the autopsy table, as it enabled him to address the context of the death so central to the concerns of the coroner's court. Indeed, it was his very involvement with the broad story line of death that made him the appropriate linchpin for medical testimony at inquests. This was the underlying rationale for the complaint addressed to Troutbeck by Ernest R. Badcock, a Battersea doctor who had been passed over as a witness in a 1903 inquest into the death of one of his patients. "Surely it is of importance in the interests of justice," he implored Troutbeck to acknowledge, "that the medical man who is first called to the case be called at the inquest. It was a frequent question of your much respected predecessor, the late Mr. Braxton Hicks, 'And, doctor, what do you know of the character of these people?' How is it possible that a stranger coming in can further the ends of justice in this way, however much he may be able to teach the general medical practitioner about his work?"[74]

Badcock's plea underscored a key point made by proponents of a practitioner-based inquest system: that inquests were not essentially about pathologically precise causes of death and that, if they were to be treated as such, they would betray their public in several important respects. First, in an argument recalling the vivid imagery conjured by opponents of Wakley's coronership over a half-century earlier, the specter of an expert-centered inquest provoked warnings of a public sacrificed to an intrusive and self-serving professional curiosity. J. Brooke Little, a noted authority on the law and history of inquests, predicted in testimony before the 1909 Parliamentary Committee on Coroners that, if inquest postmortems were subjected to a regime of expertise like that proposed by the PCC—one based on preliminary specialist examinations—"the coroner might sometimes order a post-mortem which he would not do if he had to come before the public and have a public inquiry." Asked by a committee member whether he thought there was "a danger of the coroner ordering a post-mortem for what you may call purely scientific reasons as opposed to public reasons?" Little's answer was a simple "Yes."[75]

An overly developed system of expertise promised not only to violate public sensibility in the interest of an essentially private scientific agenda, but ultimately to negate the true purpose of this form of public inquiry. Wynne Westcott, the medical coroner for the North Western district of London, while conceding the experts' claims that "pathological anatomy becomes every year a more exact science," insisted that this did not justify a radical shift in the source and shape of medical evidence at inquests. "In the ordinary course of events," Westcott explained, "the coroner does not need the accuracy of detail desirable in a hospital student's demonstration, but only a fair and honest statement of the gross appearances presented by organs, and a clear statement of the presence or absence of recognized diseases, poisons, or injuries."[76] Thus, the pro-expert contention that the testimony of special pathologists would produce more precise medical evidence was dismissed by Westcott as possibly true but, in the end, irrelevant. Earl Russell, a barrister and mainstay of the society, reached similar conclusions some years later. In his view, the inquest was a "fishing inquiry" intended to get to the bottom of a "mystery," a mystery that was not limited to the specific pathological cause of death but to a broader set of desires to know and neutralize the death, to "allay public apprehensions."[77]

The medicalization of the modern inquest, as this chapter has shown, was problematic even on its own terms, at the very place where its conceptual and rhetorical strengths ostensibly lay. What one might expect to have been a relatively straightforward process whereby modern medical expertise, using the specialist postmortem as its point of entry, encroached upon and gained priority over inquest proceedings turned into a far more opaque discussion about delineating the proper form and content of medical expertise appropriate for inquests. Framed in these latter terms, the postmortem question turned as much on debates about the past, present, and future shape of the inquest as on an internally generated discourse about the inexorable growth of expert knowledge. Medical expertise, defined by the ideals of detached universalism, promised one version of an updated, scientifically purified inquest that banished all nonproductive forms of curiosity (though, as seen in the case of John Troutbeck, not necessarily one that renounced its ostensibly ancient pedigree). A local and contextually based notion of authoritative medical evidence underpinned an alternative inquest that, its supporters maintained, might equally serve the needs of a modern polity. Here a satisfactory account of death meant a compromise between the technical precision expected of a state-sponsored scientific regime of death and the public's narrative expectations stimulated by a troubling death, expectations that remained unsatisfied by the hermetic positivism of the pathologist, his privileged access to the whispers of the dead notwithstanding. The choice of medical witness at inquests, then, was ultimately as much about a choice between versions of inquests as between modes of reading bodies of evidence.

FATAL EXPOSURES

Anesthetic Death and the Limits of Public Inquiry

Joel Simpson was scared to death of chloroform, his widow told a Southwark inquest jury summoned to inquire into the cause of the forty-four-year-old laborer's death in February 1884. According to Mrs. Simpson's deposition, her husband had gone to Guy's Hospital seeking relief from a painful case of piles and offered no objection when the admitting surgeon suggested an operation for their removal. "Deceased said to me that he did not mind what operation was performed so long as they did not chloroform him," Mrs. Simpson stated, adding that "he had an impression that if he was put under chloroform he would never get out of it." Simpson died within minutes of being chloroformed, according to Frederick Eastes, the Guy's house surgeon who administered the vapor to the "rather nervous" patient. A postmortem found, along with three large piles, a distinctly fatty heart and signs of chronic alcohol abuse. The inquest concluded with a verdict "that the deceased died from the effects of the administration of chloroform which was carefully and properly administered."[1]

The component elements of Simpson's case—fear, chloroform, death and its complex of causes, and the coroner's inquest—mark the terrain of this final chapter, which probes the limits of the accommodation sought over the course of the nineteenth century between medicine and an inquisitive public. The publicity-based ideology of the inquest, subscribed to by lay and medical observer alike, held that the dissemination of information surrounding a troubling death was ultimately a good; that secrecy wrought consequences far worse than any that might stem from overzealous exposure; in short, that an informed and engaged public was a healthy public. From this perspective, medical reformers saw their profession's role at inquests as that of an (often frustrated) adjunct to an (as yet) inadequately conceptualized, but nonetheless important, civic tribunal. Placing medical testimony on a surer footing, they insisted, would lead to a more perfect harmony between the public and scientific functions of the inquest, with the increased prominence of medicine at inquests representing an improvement in a salutary transparency of death before its public.

Here reformers assumed a specific and stable position for medicine within inquest inquiries, as an adjudicator of cases of ambiguous fatality occurring in a world outside its own doors. Social, economic, and institutional arrangements, insofar as they might figure in the causal narrative demanded at inquests, were all potential objects of a detached medical scrutiny acting in the public interest. But at the turn of century, arguments for a rigorous medically assisted public inquiry into death were made the subject of a telling reversal, when a regular procedure carried out within the confines of established medical practice was itself placed under the systematic review of the coroner's inquest.[2] Between the 1880s and the 1920s, as a consequence, the cause of inquest reform was complicated by the task of squaring theoretical support for (and participation in) the inquest's public functions with attempts to shield modern, scientific medicine from the glare of publicity. The catalyst for this controversy over medicine as an inquest object was the question about the nature and status of deaths taking place under the influence of anesthesia.

In 1896, English inhalation anesthesia celebrated its fiftieth year of service to a suffering public, a milestone that gave its leading practitioners a chance to reflect on a decidedly mixed record of achievement.[3] On the one hand, the introduction of ether (1846) and chloroform (1847) into the surgical theater was hailed as an essential contributor to the modern revolution in surgical practice, transforming operations from rough and perilous interventions into tranquil and precise ones. On the other hand, anesthesia, particularly chloroform, almost from the outset was acknowledged to be a dangerous agent of mercy. The first publicly recorded fatality from chloroform occurred within months of James Simpson's November

1847 inaugural administration in London, when Hannah Greener's death was inquired into by a Newcastle coroner's jury in January 1848. On being informed by the distinguished local surgeon Sir John Fife that "no human foresight, no human knowledge, no degree of science, could have forewarned any man against the use of chloroform in this case," the jurors returned a verdict of accidental death.[4]

Other names soon joined Greener's on the roll of chloroform victims, though in its first few decades the numbers of recorded anesthetic fatalities remained steady and fairly low.[5] A member of the Medical Society of London, Dr. Crisp, put the number of chloroform deaths between 1847 and 1852 at thirteen, while the pioneering anesthetist and celebrated public health researcher John Snow counted eighteen for the same period. Mortality figures were widely recognized as problematic, however, largely because of lax and uncoordinated provisions for reporting and registering such deaths.[6] Agreement on the actual rates of death was even more difficult to reach, especially since many hospitals did not keep a record of anesthetic administrations in their surgical theaters. As Roger Williams of Saint Bartholomew's Hospital complained to the *Lancet* in 1890, anesthetists lacked "really reliable information as to the relative frequency of such occurrences. Rough estimates there are in abundance," he observed forlornly, "the value of which may be judged from the fact that they vary, for chloroform, from 1 in 36,500 to 1 in 2666 administrations."[7]

Yet it was not the absolute number or the relative frequency of deaths under anesthesia that occupied the attention of the medical profession so much as the fact that the phenomenon was recurrent and afforded no clear causal explanation. Debates raged concerning not only the relative dangers of ether and chloroform, but also the proper mode of administration and the existence of a variety of physiological states among patients themselves (ranging from hidden constitutional abnormalities to acquired debilities attributed to causes like excessive alcohol intake) that might predetermine any fatalities. The question of anesthetic death formed the subject of inquiry for numerous medical and parliamentary commissions from the 1860s onward, but their investigations produced no clear con-sensus, leaving the question of anesthetic death at the turn of the century largely open.

Meanwhile, the death toll mounted.[8] According to the registrar general's annual returns, the 1890s witnessed a sharp increase in the number of anesthetic fatalities, which intensified discussion and dis-satisfaction with the existing state of knowledge on the subject. The fiftieth anniversary of the introduction of chloroform witnessed a sharp and "inexplicable" peak in fatalities, prompting much open-ended and at times pained reflection in medical circles about the future of the practice.[9] The *BMJ* of 12 March 1898 carried a lengthy letter under the banner "Deaths under Anaesthesia," which dolefully observed: "The jubilee year of chloroform, after fifty years' experi-ence of its use, has gone out with a record of 96 inquests held in Eng-land alone in cases of death in persons of both sexes and of all ages. This is the largest number of deaths from anaesthetics reported for any one year," the journal noted, which only confirmed its fears that "instead of improving the method of administration, the work of anaesthetics has greatly deteriorated."[10]

These figures were particularly frustrating for contemporary ob-servers because they seemed to belie efforts made in the last decade of the century by leading anesthetists to professionalize anesthetic practice. The first comprehensive textbook devoted entirely to the subject, Dudley Buxton's *Anaesthetics: Their Uses and Administra-tion,* appeared in 1888, and within five years J. F. W. Silk's review of the twelve metropolitan medical schools found that all but one had a special anesthetist attached to it; nine of these schools included instruction in anesthesia in their curricular prospectus.[11] The 1890s witnessed the founding of the first society of anesthetists (1893) and the publication of the first specialist journal (1898), both of which actively promoted their specialty as an integral and regular part of the medical profession and sought better conditions of training, ex-amination, and research.[12] Despite these developments, to be sure, anesthetics was still very much in its infancy. Instruction remained patchy; for example, when Silk looked more closely at the actual practices of metropolitan medical schools, he concluded that only 35 percent of their graduates could be described as having been given adequate "special" instruction, a rate that dropped to 18 percent when the rest of the United Kingdom was considered.[13]

Anesthesia in England around the turn of the century, then, was in the ambiguous position of being—at least in theory—more institutionally grounded than ever before, but at the same time vulnerable to the charge that, a half-century of experience notwithstanding, it remained a dangerous and unreliable practice. Leading figures in the field, while maintaining that the number of fatalities was in reality low, especially considering the proportional increase in dangerous surgical operations themselves made possible in large part by anesthetics, acknowledged that they faced severe problems of public mistrust. Even as the investigations, debates, and reforms continued to occupy the medical profession's attention, at the end of the century these were joined, and in some sense overtaken by, concerns of an entirely different nature. These involved questions about the proper framework within which anesthetic deaths ought to be investigated. At the center of this controversy stood the coroner's inquest.

I

For coroners, anesthetic deaths occupied a special place among the range of fatalities involving recognized medical intervention. Most of his colleagues, the secretary of the Coroners' Society asserted in 1894, made a practice of distinguishing between deaths following hospital treatment in general and hospital deaths involving anesthesia. Coroners typically involved themselves in the former cases only if special circumstances warranted (if, for instance, surviving relatives requested further inquiry). However, he explained, "it is a general rule to hold inquests in all cases of death occurring while under the influence of anaesthetics."[14] Ten years later the society sought to translate accepted practice into a more explicit policy of differentiation. Its council recommended that

> inquests should be held in all deaths occurring whilst under the influence of an anaesthetic, irrespective of whether the friends and those specially concerned in the administration of the anaesthetic and the operation were satisfied in every way. In cases, however, where the patient had

recovered from the effects of the anaesthesia, and the death was due to a necessary operation or the disease for which the operation was performed, an inquest is not required, provided always that there is no allegation of neglect, serious rumours, or presumed want of skill on the part of the operator, or that the friends of the deceased do not request an inquest for their satisfaction.[15]

The society thus explicitly isolated anesthesia as a uniquely disturbing element in the complex of surgical practice. Surgical deaths were not worthy of inquiry in and of themselves, requiring an inquest only to the extent that the circumstances generated rumor, suspicion, or other potentially disruptive reaction. Anesthetic deaths, hitherto ambiguously distinguished from surgical deaths, were by the council's recommendation to be placed in a category all their own.

Individual coroners gave different accounts of why anesthetic death ought to be subject to special scrutiny. Some, like E. A. Gibson of Manchester, held that inquests were necessary by virtue of the toxicity of the anesthetic agent. When a medical man anesthetizes a patient, Gibson asserted in testimony before the 1909 Parliamentary Departmental Committee on Coroners, "he is poisoning him . . . to the verge of safety. If the patient dies from the anaesthetic, the man who has administered the poison has killed him, and, therefore, in a sense, it is for him to justify what he has done."[16] Others held that the use of anesthetics in the surgical theater rendered the patient peculiarly unable to exercise normal care and control over his wellbeing. Those who took this position, like Birmingham coroner Isaac Bradley, drew an analogy between prison inquests and anesthetic deaths taking place in the operating room. Prison inquests were necessary, Bradley explained to members of the same parliamentary committee, "the obvious reason being that the prisoner in the prison is not a free man—he is under restraint; he is not free to come and go . . . Similarly, as I think, a patient who has been anaesthetized is not free to come and go; the patient has been deprived by artificial means of his volition."[17] According to John Troutbeck, the case for inquests into anesthetic death—and perhaps into surgical deaths more generally—was simpler still: These were "violent" deaths, and thus commanded the coroner's attention.[18]

Anesthetists reacted to such suggestions with frustrated indignation, as the association of anesthesia with violent or otherwise suspicious deaths seemed to ignore the central fact of the anesthetic revolution. Debates on its relative safety notwithstanding, anesthesia was widely touted as one of the major developments in the recent history of medicine, responsible for transforming the scene of surgery from one of disruptive pain to one of beneficent quiescence. An 1851 *Lancet* editorial gave its impressions of this "valuable boon to suffering humanity" in terms that would soon become standard fare in describing the effects of anesthetics on surgical practice:

> The knife of the surgeon lost, as it were, by magic, all its terrors. When the sufferer for the first time was presented to the eye of the spectator, lying passive under the influence of chloroform, how strongly was marked out the difference between the sensible and the insensible object of the operative procedure. There was no longer witnessed the cry of agony issuing from the frail body of some poor emaciated woman, whose breast was about to be submitted to the knife; nor the scarcely less painful effect of subdued emotion, in the strong frame, while it quivered under the strokes of the scalpel. The surgeon now has not to contend against these calls upon his humanity . . . There lies the patient, under the influence of the Lethean vapour, reveling perhaps in dreams of happiness, whilst the operator is employed in removing a limb, or dragging away some portion of necrosed bone,—the patient not being the least sensible of either the pain or the danger of the operation.[19]

Anesthesia's "magic" lay in its capacity to turn oblivion, insensibility, delusion—and the resulting asymmetrical subjective competencies of the desensitized surgical theater—to common advantage. Formerly, the encounter between surgeon and patient as sentient beings made for a brutalizing scene—for the patient through his or her experience of pain, for the surgeon through the contradiction between the immediate physical consequences of his actions and their ultimate therapeutic intent. In the new scenario a humane surgery was the outcome of a salutary dehumanization: an objectified patient, insensible both to the physical assaults made on the body and to the possible negative consequences of this invasion, and a surgeon who, freed from the complications of pathos, was capable of skillful fidelity to the knife's clarity of purpose.

At the fiftieth anniversary of the advent of inhalation anesthesia, F. W. Hewitt, the leading practitioner in the field and author of a celebrated textbook on anesthetics, reaffirmed its revolutionary influence on medicine and extended the liberating effects of a redistributed sensibility beyond the confines of the surgical theater:

> By the assistance which Anaesthesia had rendered, countless advances and developments in surgical science have resulted. Moreover, Anaesthesia has calmed the public mind. In past years the services of her master were only requisitioned as a last resource; they are now as often invoked in doubtful or obscure cases, and in those in which inconvenience rather than actual suffering has to be remedied. By her influence, too, the very nature and disposition of her master have been modified and softened. The constant infliction of pain to which he had grown accustomed tended to dull the edges of natural sympathy, and to bring about in him an apparent if not true heartlessness of demeanor and disposition. All this has been changed by Anaesthesia.[20]

Anesthesia was a "faithful handmaid of her master Surgery," one that not merely assisted but transformed. By rendering the patient a pliant surgical object, it advanced scientific medicine by making protracted and innovative operations possible. Furthermore, the promise of insensible objectification encouraged on the part of the suffering public a willing submission to the surgeon's knife, turning the operation theater into a sought-after therapeutic site rather than a terrible refuge of last resort. Through the medium of anesthetic vapors, finally, the surgeon's own professionally acquired insensitivity—and perhaps more importantly his public reputation for insensitivity—could be transposed onto his surgical object. To class a death under these circumstances as a potential violation worthy of public inquiry was as illogical as it was ungrateful. H. Bellemy Gardner, a practicing London anesthetist and an ardent defender of his subspecialty, spoke for his fellow practitioners when he insisted that "a death in a beneficent unconsciousness cannot be held to be 'violent.'"[21]

In these tranquil depictions of the anaesthetized surgical theater, however, there was scope for seeing not an absolute banishment, but a mere displacement of pain and violence, one that carried with it its own potential for abuse. By rendering the patient a pure surgical object, anesthesia weakened the constraints placed on operative practice. The anesthetized patient could neither through a conscious act of will nor through the corporeal language of pain resist the surgeon's actions; the surgeon, on the other hand, might view the patient as mere object, the surgical scene evacuated of the beneficial limitations imposed by human empathy. This concern was never far from its counterpart, at times explicitly invoked as a qualification of anesthetic "magic." Indeed, the *Lancet*'s early declaration of awe at the power of anesthesia also expressed reservations about "the facility with which patients are now persuaded to submit to the knife, and the encouragement which it holds out to what are called 'promising young men' to 'carve their way into practice,'" the banishment of pain enabling them to excise offending bodily matter "with as much nonchalance as though it were being removed from the dead body in the dissecting-room."[22]

The disruption of the anesthetic ideal was also an implicit feature of the professional discourse of anesthesia. Discussions of the anesthetist's responsibilities, like that in the "Medico-Legal" section of Hewitt's classic textbook (composed by the barrister Digby Coates-Preedy), were deeply concerned with broad questions of rights, consciousness, and will, their possible violation in the context of administration, and their implications for anesthesia as a matter of public interest.[23] Anesthetization without the consent of a mentally competent adult or guardian constituted a clear case of assault, Cotes-Preedy asserted. The situation was more complicated, however, when the patient, having given prior consent, changed his or her mind at the moment of anesthetization.

> It sometimes happens that an excitable patient who has consented to an operation, and to anaesthesia as a part of it, changes his mind when brought to the operating table and refuses to accept the administration. In such an event the anaesthetist may endeavor by argument to convince the patient and so persuade him to submit to anaesthetisation. But he may not forcibly restrain the patient and then proceed to anaesthetise

him. On the other hand, if a patient accepts the administration, and when partly anaesthetised struggles to tear the mask from his face and to interrupt the anesthetisation, the anaesthetist is justified in restraining his struggles and forcibly completing the induction, if satisfied that the patient's powers of reason and judgment are sufficiently suspended to render him irresponsible for his action.[24]

By proposing this scenario Coates-Preedy constituted the administration of anesthesia as a procedure suspending the normal rules governing the interaction between individual subjects: self-control on the part of the patient is alienated (at some ill-defined point determined by the anesthetist) as an integral component of the transaction. With the power of determining the patient's rational capacity arrogated to himself, the anesthetist becomes the temporary custodian of the patient's prerogatives as a rights-bearing subject and thus becomes himself a potential subject for the scrutiny of an inquest jury.

Nor, finally, was the vocabulary of violence—assault, force, restraint, struggle—limited to theoretical discussions of responsibility. In textbooks, in journal articles, and in testimony at inquests, anesthetists themselves commonly and unselfconsciously noted that the patient "struggled violently," "took the anaesthetic badly," or "had to be held down by the Dressers." Dudley Buxton's entry on anesthesia in the 1899 edition of the *Encyclopaedia Medica* considered this common enough to constitute its own recognized part of normal anesthetic administration, the "Stage of Excitement," in which patients exhibit "a boisterous excitement, attempt to sit up on the table, to wrestle, fight, and so on—speaking, entreating, remonstrating, threatening, singing, praying, even cursing." When passing through this stage, Buxton advised, a patient "should be restrained, one assistant leaning over the knees, another holding a wrist in each hand, while the chloroformist keeps the head down and prevents the patient sitting up, by the pressure of his left hand upon the brow . . . Despite what has been said, no harm comes of firm restraint, avoiding of course the appearance or reality of roughness."[25] Through such reports from the operating room, anesthetists themselves reintroduced the open, provisional, unfinished elements of surgery ostensibly banished along with suffering. Amid the wrestling, praying, and restraining, an observer might well conclude that some form of outside scrutiny was in order.

Indications from within the surgical theater that anesthesia had not induced an unambiguous calm intersected with a broad set of contemporary concerns about the treatment of patients by a hospital-based medicine that seemed to be increasingly dominated by values of pure science. Of the numerous late Victorian campaigns publicly challenging this ostensibly new and inverted relationship between healer and patient, it was the formidable antivivisectionist movement that is most important for our purposes, because of its high visibility and its consequent influence on the broad discursive field within which the more narrowly defined question of anesthetic inquests was framed.[26] Reaching out to related campaigns against an authoritarian and exploitative medical regime—notably those contesting the Contagious Diseases Acts and compulsory vaccination legislation—as well as to a wider network of contemporary moralizing social crusades, the antivivisection movement mounted a highly public challenge to the contemporary embrace of experimentalism. Its centerpiece, of course, was the suffering entailed in animal experimentation and the dehumanizing effects of such activities on those undertaking such work, but the movement also attacked anywhere it saw the cure of the patient being subordinated to the collection of knowledge.

Hospitals came under close scrutiny, with antivivisectionists seeking to demonstrate the fact and the implications of a fundamental split between an "old" humane and a "new" scientific school of medicine. While the former seeks out quick and painless cure, the latter, in the words of the *Daily Chronicle,* views the patient not so much as "a human being who has come into a hospital to be cured and sent home at the earliest possible moment, but as a very interesting composition of flesh, bones, and blood, from which some instruction is to be derived; some new discovery made which will entitle the discoverer to read a paper before a medical society or to write a letter to the *Lancet* or the *British Medical Journal.*"[27]

This specter of human experimentation served as a potent and palpable image in the antivivisectionist armory, one that had clear implications for the debate on anesthetic death. Thus, while anesthetists like Hewitt enthused about the reflowering of surgical humanitarianism that followed the operator's release from the deadening consequences of visiting pain upon a fellow creature, critics saw only a self-serving callousness filling the space left by the elimination of some form of mutually shared suffering, however asymmetrical in its nature and effects. In 1898 *The Nineteenth Century,* which over the past decade had served frequently as a forum for debating the relationships among medicine, science, and society, ran an exchange between the antivivisectionist Elizabeth Moss King and the anesthetist Dudley Buxton. King's initial article, "Deaths under Chloroform," related her own terrifying experience of being given chloroform by what she and others called the "smothering" method. *Smothering*—that is, covering the face with a chloroform-saturated cloth—was the quickest and cheapest method of chloroforming, but as the term suggested, it was also thought to be least tolerable for the patient. Habitual recourse to this method by the "School of Stiflers" served for King as a signal example of the tendency in modern medicine to sacrifice the patient to medical expediency.[28]

King subsequently enlarged upon her attack on anesthesia's complicity in the trend toward a detached and objectifying medical ethos in "Death and Torture under Chloroform." One of her medical correspondents, King reported, told her of being

> painfully struck with the callousness shown in chloroforming patients in the operating theatre at hospitals. The wretched patient, highly nervous at the prospect of an operation and the mysterious terrors of chloroform, has to face all the young students in the theatre, who are watching his agony of fear with far more curiosity than sympathy. To them he is only a "case." A natural instinct prompts every animal, whether brute or human, to hide himself when in pain or grief, and this exposure to a crowd of callous, curious strangers at such a crisis must be a very terrible ordeal to go through.[29]

Here King's critique utilized the most powerful trope of the "scientist at the bedside" against the anesthetist, charging him with complicity in the betrayal of a humanitarian therapeutics to the interests of modern experimentalism practiced by a rising hospital elite.

in its details but misleading and thereby debilitating in its
effect.[37] In turning matters of science into journalistic comes, newspapers were not merely exposing matters better left
from view, they were distorting truth and stimulating mor-
session.

Lancet no doubt had a point. Take, for example, the fol-
account of a 1907 London inquest appearing in the most
read newspaper of the period, *Lloyd's Weekly:* "LONDON
STS. THIRTY-EIGHT DEAD. High Anaesthetic Death Roll at
Leads to Inquiry."[38] Only later in the report did it become clear
he figure referred to deaths over the last six and a half years.
was precisely the sort of "compressed" reporting that drew the
f the Guy's surgeon, Claude Miller, at a Southwark inquest.
d by the coroner whether inquiry into anesthetic death ought
e public, Miller lashed out against the sensation-mongering
cs of the popular press, for whom, he complained, "Another
dal at Guy's' made a capital headline."[39] Such exposure, Miller
ed, had serious practical consequences beyond alarming the
ing public. He told of a recent case in which a critically injured
was nearly refused admittance: "An operation was found nec-
ry, but before it was undertaken the doctors discussed the mat-
because the deceased was so bad that they thought that if he died
e he was under the influence of an anaesthetic the newspapers
uld make a fuss of it."[40] The type of exposure dictated by the
eric requirements of a popular journalism feeding off the mate-
generated by public inquiry was in Miller's view directly inimi-
to the interests of the individual patient and the public at large.
Complaints that the harassment of anesthetists by untrammeled
blicity damaged reputations, retarded progress in the discipline,
d unnecessarily alarmed the public found a wide echo. "There
pears to be a growing fear of anaesthetics among our medical
dents," a 1910 *Lancet* editorial worried, "and some would have
believe that the dread of a *début* made in the coroner's court
ids to aggravate this fear rather than to stimulate the student's
thusiasm for increased knowledge of the practice of anaesthesia.
this is so the dread engendered in the public mind by reading a
rbled account of inquests offered for their perusal in the daily

Against this trend, antivivisectionists insisted, the patient had only
one defense—public scrutiny. It was on this principle that the Society for the Protection of Hospital Patients was founded in 1897; its
primary aim was the reversal of the "growing tendency" of hospitals to regard themselves as "endowed schools for experimental research." Modern surgery, the society bluntly charged, indulged a
"habit of experimenting on patients for purposes other than those
of their own immediate benefit or relief."[30] The antivivisectionist
journal the *Verulam Review,* commenting on the prospects for the
society's success, advised it to take heed of a fundamental truth: "For
ourselves we would suggest to all interested in the question one safe-
guard upon which they would do well to insist most strongly. It is
publicity."[31]

The medical profession's most common response to charges of
human vivisection was one of studied neglect, seeking, it would seem,
to minimize direct engagement lest it should seem to lend credibility to outrageous calumny. Yet this strategy of isolation was not
always deemed possible or even expedient. The very visibility of
King's critique drew Buxton into a reluctant, and paradoxical, journalistic dialogue in an attempt to uphold his profession's injunction
against "ventilating medical questions in non-professional journals."[32] Moreover, as the *BMJ* lamented in reaction to a recent directive issued by the London County Council (LCC), the charges
leveled against hospital practice met with occasional success and,
thus, had to be acknowledged and forthrightly confronted. In its
1895 report on London inquest practice, the LCC recommended
that every case of death after surgical operation should be reported
to the coroner. In condemning the conceptual basis upon which such
an inexplicably sweeping proposal rested, the *BMJ* named the suspected culprit: "It may be that there is an uneasy suspicion amongst
a portion of the public—excited and fostered by antivivisectionists,
antivaccinators, and the like—that operations are done in hospitals
for the sake of experiment." Having made this unhappy association,
the editorial reluctantly conceded the contingent benefits of such a
policy: "It may be politic to allay all genuine uneasiness by offering
every possible facility for investigation in such cases."[33]

II

Politic though it may have been, the exhortation "to allay all genuine uneasiness" was difficult to realize. Fear, anesthetists acknowledged, was their constant companion and was only partly subject to their control. Leading voices in the field urged practitioners to be more conscious of their own part in abetting uneasiness, advising modifications in protocols of administration with the patient's subjective state in mind. "The majority of patients regard the anaesthetic with far greater dread than the operation," the celebrated surgeon Frederick Treves observed. "Of the surgeon's work they are assured they will know nothing; but they do know that they will be horribly conscious of those palpitating moments which precede the onset of the gruesome and unholy sleep." It was the duty of the anesthetist to take active steps to counter this disturbing image, to consider, for example, the way he approached his enervated patient, for whom the anesthetist's mask appears as "a symbol of the Valley of the Shadow of Death." Treves counseled a posture of humane simplicity, condemning both the "administrator who solemnly displays a copious apparatus which he manipulates with the stolid ostentation of an executioner" and his polar opposite, "the anaesthetist who calls jauntily for a folded handkerchief, and, after placing it over the patient's trembling face, proceeds to chatter incontinently of his summer holiday."[34]

Anesthetists felt they were making progress in their battle with the sources of fear identified by Treves. In the introduction of the 1912 edition of his textbook, Hewitt praised the care taken in recent years to have the patient "spared, as far as possible, all visual and aural distress immediately prior to and during anaesthetisation" and did his part to encourage the trend by including a section devoted to "Administration from the Patient's Point of View."[35] But in their attempts to instill a rational calm in patients, anesthetists and their allies felt themselves fighting a losing battle, thwarted by the pervasive and insidious influence of the mass press. "All the narratives of 'death under anaesthetics' which are detailed at coroners' courts find their way into the daily press," the *Lancet* complained in 1914. "The reports are compressed into a few lines, and almost invariably the death appears as if it had been caused by the anaes-

Into the Valley of the Shadow of Death. J. T. Clover (ratus for administering chloroform, c. 1880. Courte Library, London.

thetic. These newspaper paragraphs recur day sick, and those whose lives depend upon a succe ing undertaken, become obsessed with an unr taking an anaesthetic."[36] This was a problem (*Lancet*'s estimation: the newspapers in the age of ism" produced, as a generic necessity, an accou papers is reacting in a sinister manner upon medical training."[41] The fact that public reporting on anesthetic death also made patients reluctant to submit themselves for necessary operations, the journal noted a few years later, made the problem "doubly dangerous; on the one hand, it may lead the patient to refuse treatment . . . ; on the other, it may throw anaesthetists into a state of panic, thus engendering the worst state possible for successful and safe anaesthesia."[42] A 1913 draft memorial prepared by the Royal Society of Medicine's Section on Anaesthetics to protest the excessive public scrutiny of its members' "failures" stated the anesthetists' complaints with an exasperated clarity: "We submit that no other kind of professional career could be carried on and survive the publication of its failures alone, for the veritable breath of life to a barrister, professor, or doctor, is a reputation for success."[43]

Publicity thus worked against the successful outcome of modern surgery at a surface level, disrupting what was optimally a confident, professionalized interaction among surgeon, anesthetist, and patient. But publicity worked against anesthesia in a still more direct and essential way: The public's "obsession" induced by garbled reports taken from anesthetic inquests was in itself pathogenic, a direct contributor to the physiology of anesthetic fatality. The "anaesthetic terror," in short, was itself lethal.[44]

It had long been suggested that the patient's mental state might be relevant to the outcome of anesthetic administration, but it was several decades into the British experience with inhalation anesthesia before the inquest itself became a feature of an explicit pathology of fear. In 1870 the *Lancet* proposed an early formulation. In taking note of a recent spate of press reports of deaths under anesthesia, it observed that, "although these so-called 'accidents' bear an infinitely small proportion to the cases in which the most popular of anaesthetic agents is administered without injury, yet they serve to alarm patients and their friends, to surround the idea of an operation with unnecessary anticipations of evil, and possibly, in some cases, to modify through the emotions the ultimate results of treatment."[45]

By the first decades of the twentieth century, this connection had become a common weapon in the anti-inquest arsenal. At a 1908 discussion of anesthetic fatalities, for instance, members of the

Medico-Legal Society were quite confident in making the association. J. F. W. Silk, an anesthetist at several of the London hospitals, stated that "one of the favouring factors in causing these deaths was the state of terror which was often induced in a patient by reading sensational literature on the subject."[46] The celebrated professor of forensic medicine at Glasgow, John Glaister, agreed, advising restraint in public discussion of anesthetic deaths because "he believed that the factor of fear had itself something to do with the fatal result."[47] In testimony before the parliamentary committee on coroners a year later, the world-renowned surgeon and unapologetic vivisectionist, Sir Victor Horsley, remarked rather casually that "these inquests are simply a source of alarm to the public and make them nervous about taking anaesthetics, and contribute thereby to heart failure and so forth by fainting."[48] Anesthetists as a professional body incorporated the notion of fatal publicity in a 1913 protest memorial to coroners: "We frequently find that patients about to undergo surgical operations are reduced to such a condition of emotional shock from fear of the anaesthetic as a result of these reports that the danger of their collapse in anaesthesia is much increased."[49]

The trauma of pain experienced in the preanesthetic era, according to these analyses, had been reintroduced in the guise of fear. The link between fear, publicity, and death was the most essential argument against inquests into anesthetic deaths. The claim that public knowledge about the dangers of anesthesia was in itself lethal brought the requirements of hospital medicine into direct and irreconcilable conflict with the logic of the public inquest, according to which society functioned best when the mysteries of death were dispelled under the pure light of public scrutiny. In this instance the inquest intensified rather than dissipated anxiety because the jury's tendency to an alarmist empathy with the deceased precluded its normal capacity for rational judgment. Nor was this state of perverted judgment limited to the jury. An invidious and misguided regime of exposure had broad implications for the well-being of the social body writ large. In arguing against public scrutiny, the public figured not as the embodiment of common sense, but as a collective of prospective and vulnerable patients, able neither to judge nor even to read about anesthetic deaths from a position of reasoned detachment, ever liable to imagine—and thereby position—itself as anesthesia's next victim.[50]

Critics of inquests into anesthetic deaths were by no means opposed to all forms of inquiry, however. Anesthetists and their supporters consistently urged the desirability of and, indeed, the necessity for coordinated and purposeful investigation of the phenomenon, that is, scientific inquiry. On the surface this alternative was straightforward, entailing the systematic correlation of clinical and postmortem information over the spectrum of anesthetic deaths for the purpose of improving knowledge and practice. Analysis of the discursive framing of the proposed "scientific inquiry," however, suggests that at another level the very notion of what was "scientific" about such inquiry was informed by the contested standing of the inquest and its "public" and, furthermore, that publicity was not the irreducible nemesis of the scientific, but within proper limits could function as a useful and necessary supplement.

The hearings held before the 1909 House of Commons Departmental Committee on Coroners provided a prominent stage for a systematic consideration of the relationship between these two forms of inquiry. The committee had as one of its explicit terms of inquiry the vexed question of anesthetic inquests, and members accordingly questioned leading anesthetists, pathologists, and surgeons on the relative merits of public anesthetic inquests as compared with alternative proposals for making investigations more scientific—to appoint expert assessors to assist coroners, to dispense with inquest jurors, and, more radically still, to replace public inquiry itself with an internal investigation conducted by hospital staff and government experts. Witnesses like Buxton and Hewitt agreed that the coroner should sit with an expert assessor and that juries were at best incompetent to deliberate on questions that were to them, in Buxton's words, "Absolute Greek."[51] Most cases of anesthetic death, Buxton further explained, involved clear issues that might be decided by informal (i.e., internal) inquiry with the assistance of hospital personnel and facilities. Such a change in procedure would provide rapid and accurate determination of the medical cause of death, avoid wasting the time and public funds required for full inquests, and finally curb much of the press coverage of these deaths which, in Buxton's estimation, was "undoubtedly a danger to the public."[52] Hewitt concurred, counting among the benefits of a more limited medical investigation into anesthetic deaths the fact that they would "not be quite so public; that is to say there would be less chance of the Press taking up these cases and alarming the public by their publication," and that "the evidence obtained would be much more serviceable, much more valuable, and much more scientific."[53] In the view of anesthesia's leading lights, clearly, the defining other to *scientific* was *public*.

The committee took these views into account in its "Report of Inquiry into the Question of Deaths Resulting from the Administration of Anaesthetics." This document charted a careful course between the demands of transparency and of science, recommending that, while coroners should continue to be informed of all anesthetic deaths, they should exercise greater discretion—and restraint—in determining whether to hold a full inquest. Two further proposals were designed to place the issue of anesthetic death within a more stable and productive framework: first, in the event of an anesthetic death within any hospital or similar public institution, the authorities of the institution should hold a "scientific investigation into the actual cause of death"; second, a small standing committee on anesthetics should be instituted under the authority of the Home Office to collect and digest information relating to anesthesia, to report on advances and discoveries in the field, and to direct chemical, toxicological, and physiological research. The implication of the committee's recommendations was that, while the inquest should be preserved in its demonstrative function, the instances of its demonstrations ought to be chosen with greater care and economy, with the internal hospital investigation taking over as the preferred method of establishing a more constrained, functional, and ultimately more palatable significance of anesthetic death.

Even at this moderate level, however, the shift to a more scientifically based inquiry proved problematic, as illustrated by the attempt to set up a pilot scheme for London in the early 1920s. The proposal, drawn up by a joint committee of representatives from the Home Office and Ministry of Health, with advice from the Medical Research Council, called for the formation of a panel of expert pathologists drawn from London hospitals who would be made available to coroners for postmortem examinations in anesthetic inquests and for the collection and analysis by the Ministry of Health of internal reports made by hospitals on anesthetic deaths occurring within their walls. But, as the committee's representative from the Medical Research Council, Sir Walter Fletcher, frequently pointed out, the scheme was mapped out over fraught terrain.[54] Fletcher's own view was that the scientific case for systematic inquiry was dubious, observing in a letter to his Ministry of Health counterpart that he doubted "the proposed collection of anaesthetic deaths . . . will help us to get new knowledge." He supported the scheme, however, on public rather than scientific grounds: "In view of the dramatic nature of these deaths, and of the way in which they get undue notice in the press," he wrote, "I do think that there is a good case for the Ministry letting it be known that they are watching these deaths closely, and receiving special reports on them." That the government should be seen to be attentive to the question of anesthetic death was a good in itself, but only up to a point. Official inquiry should signal purposeful activity, Fletcher advised, while avoiding the appearance of validating the public's unhealthy obsessions:

> I feel very strongly indeed, though, that it would be definitely against the public interest for the Ministry to announce that some special Committee had been appointed to investigate deaths from anaesthetics. I think this would simply add to the quite artificial exaggeration of present dangers, already mentioned. Taking an anaesthetic is less dangerous than walking across the Strand, and yet it is definitely known that a mere feeling of alarm on the part of the patient increases the risk, at least where chloroform is used. The Ministry should, I think, use every effort to magnify the safety of anaesthetics, and do nothing to point to risks, in the interests of the suffering public.[55]

Fletcher's painstaking remarks on the complexities involved in even considering the subject of anesthetic death on an official level testify to the entangled and volatile relationship that obtained between publicity and investigation. The proposal to assemble an expert panel of pathologists was similarly problematic, cutting as it did to the symbolic core of the inquest system—custody of the dead body. As the responses to the joint committee's questionnaire circulated to London and provincial hospitals in October 1920 made abundantly clear, the coroner's jurisdiction over the corpse was a standing irritant to proponents of contained scientific inquiry. The Middlesex Hospital complained bitterly that as a consequence of the coroner's intrusion "we have no control over the conduct of the post-mortem examination." London's University College Hospital reported that "the scientific value of [its internal investigations] is greatly minimized by the existing system which precludes post-mortem examinations in such cases being held by the Hospital Authorities." Saint Thomas's similarly charged that its efforts at investigation were "from the scientific point of view . . . rendered completely sterile by the fact that when a death occurs under an anaesthetic the body is removed to a public mortuary and the autopsy is performed by some person appointed by the coroner."[56]

For hospital officials, the inquest disrupted a hermeneutic circuit that had as its center the examination of the physical body, removing anesthetic death physically, conceptually, and rhetorically from the purposeful confines of the medical inquiry. But, as far as the Home Office representative on the committee was concerned, hospitals would have to live with this frustration. For his department to give its blessing to the idea of a panel of pathologists, Arthur Locke insisted, "it would be necessary to recognize fully that the custody and control of the body are with the Coroner, and make it clear that no person should have access to the body without the Coroner's authority."[57] The body in question was emphatically more than a unit of scientific datum.

III

Anesthetists and their allies were unified in their opinion that the inquest, as far as its capacity to determine the scientific questions posed of an anesthetic death, was at best useless and obstructionist, at worst lethal. Even its harshest medical critics acknowledged, however, that an inquest could provide an explanatory framework that a strictly scientific account could not and, further, that in certain instances this alternative model had its uses. The inquest could serve to manage the disturbance inevitably created in the wake of an anesthetic death by encasing it within a publicly recognized and recognizable code. Contained within its proper limits, the coroner's inquest could serve as a valuable adjunct to the anesthetic project.

The medical press had, of course, long acknowledged the explanatory and neutralizing potential of public inquiry, and it was easy to see how this might work in the specific case of anesthetic deaths. "The vast majority of cases reassures the relatives of the deceased and builds up public confidence," a *Lancet* editorial observed. "No tribunal of scientific experts sitting behind closed doors would carry the same weight with the multitude as the open court of the coroner," it concluded, "where gossip and suspicion are vindicated or refuted and human actions and motives are challenged and defended in an atmosphere which would be wholly uncongenial to pure scientific inquiry."[58] This demonstrative, extrascientific function of inquests into anesthetic deaths was recognized across a wide spectrum of opinion, as the testimony of witnesses before the 1909 parliamentary committee made quite clear. Victor Horsley combined his usual scorn for inquests with a recognition of their contingent utility. Asked why anesthetic deaths had been constituted as a specially sensitive category, Horsley replied: "The reason why the custom of coroners' inquiry has grown up [around anaesthetic death] is obvious, namely, that a perfectly healthy person may be suddenly put to death, and if the public wish an inquest to be held, it must be held to satisfy the public mind; but the verdicts returned, in my opinion, are perfectly ridiculous."[59]

Horsley at the same time dismissed the inquest as unedifying drivel and acknowledged its public, performative status. The bind for a man of science like himself was that the inquest produced a narrative of medically induced death that was, by his standards, wrong but that in practice achieved the desired explanatory and neutralizing effect which his own account could not deliver. Witnesses more sympathetic to the inquest system also acknowledged this feature of the coroner's inquiry. Birmingham coroner Isaac Bradley believed that, while the inquest might uncover evidence of error in administration of anesthesia, of equal if not more importance was its enactment of protection and its capacity to contain anesthetic death within a sanctioned public discourse. One of the primary objects of holding inquests into anesthetic deaths, he stated, "is to maintain public confidence in the medical profession and medical institutions; and I do think it does maintain that confidence for people to feel, when a death occurs under such circumstances, that it is inquired into."[60] When Bernard Spilsbury, a regular participant at London anesthetic inquests and the rising star of early twentieth-century forensic pathology, was asked if he considered it desirable that a postmortem be held in every case of anesthetic death, he replied: "For the sake of the anaesthetist only, it is an important circumstance," as in his experience postmortems generally cleared the anesthetist of any suspicion of wrongdoing.

In remarks made at a Royal Society of Medicine meeting several years later, Spilsbury elaborated on this observation. The necessity for anesthetic inquests, he maintained, should be determined according to the circumstances of the case. If the patient had been moribund before the operation, "no great advantage was gained by an inquest." If, on the other hand, the patient had been in good general health and not seriously ill,

> it was essential that a public inquiry should be held in the interests alike of the anaesthetist, the surgeon, the hospital, the relatives, and the general public. Few cases were more distressing to relatives who, after attending an inquest, were often satisfied that everything possible had been done, and that the cause of an unexpected death had been cleared up in the coroner's court. If no public inquiry were held the patient's friends might well retain a feeling of suspicion or resentment. An indignant parent might then write to the press, giving a perverted or exaggerated account of the circumstances. A special inquiry by a committee . . . would certainly be helpful from a scientific point of view, but would not replace the public inquiry in this class of case.[61]

Spilsbury thus envisioned a flexible set of procedures sensitive to the requirements of the situation and the audience at hand. Anesthetic deaths already encoded, for whatever reason, in the public mind as normal or otherwise sufficiently accounted for could move immediately into the sanitized realm of the scientific investigation. Those that called attention to themselves as aberrant had, ultimately for the sake of the medical profession itself, to be submitted to the ritual of the coroner's inquest. The logical extension of this argument, left unstated by Spilsbury, would be that the inquest functioned within the system of naturalizing heretofore considered abnormal death—that the more kinds of anesthetic deaths the inquest explained, the fewer kinds of anesthetic deaths were read as abnormal.

This leaves a basic question unanswered: In what respects did inquests perform a naturalizing function with respect to anesthetic deaths? Certainly, inquests were not prone to find verdicts of culpable negligence. Hospital-based anesthetic death was not mentioned in discussions of malpractice cases by any of the major contemporary works on medical jurisprudence, nor were any such examples cited in the leading law and medical journals of the day. The *Lancet,* in fact, complained of the "melancholy sameness" of inquest verdicts,[62] while the *Law Journal* advised anesthetists not to disregard the importance of reforming their procedures even though no instance of an inquest jury deciding against one of their brethren could be cited. "If the medical profession does not itself move in these [reformist] directions," the *Journal* warned, "its actions may be hastened by a verdict of manslaughter returned by an indignant jury."[63] In his 1922 review of the medicolegal aspects of anesthesia, finally, Digby Coates-Preedy mentioned only cases involving herbalists and similarly "unqualified" practitioners and observed that, because "the Courts have not been called upon as yet to lay down any ruling on the point," the legal responsibility of the anesthetist for the outcome of an operation remained ill-defined.[64]

The argument suggested here to account for this absence is not that coroners acted in deliberate collusion with hospital medicine to cover up potentially implicating deaths.[65] It is rather that the very structure and rationale of the inquest—having as its end the provision of a publicly derived supplementary narrative for deaths regarded as disruptive or disturbing—made it amenable to, even reliant upon, recognized and recognizable modes of causal explanation which, however provisional, incomplete, or avowedly fictional, worked as such. The stated function of the coroner's inquest was to establish cause of death through public inquiry. At the best of times, arrangements for carrying out this task were subjects for dispute. But the case of anesthetic deaths was of special difficulty because the question of the cause of death was itself such a central and contentious feature of the subject of anesthetics.

Broadly speaking, the debate about the causality of anesthetic death was split into two positions. One school held that such deaths were related to the conditions of administration and reception and thus positioned the anesthetist, as a skilled practitioner in control of his subject and his tools of trade, as the critical variable in determining outcome. Others emphasized the condition of the patient—notably his or her constitutional "idiosyncrasy" and preexisting organic dysfunction—as the key to tracing the roots of an aberrant death.[66] On the level of scientific research, both positions had adamant champions, though the terrain grew less polarized with the rise in the claims to professional status on the part of hospital anesthetists. As leading anesthetists began to present themselves from within recognized, secure positions in specialized journals and meetings, the voices supporting the premise (first laid down by John Snow in his early studies of chloroform and ether fatalities) that the path of progress lay in concentrating on techniques of administration drowned out those insisting on the structural necessity of anesthetic death. As a result, a type of agnostic truce in the conflict over the causal explanation of anesthetic death prevailed by the turn of the century. Researchers and practitioners agreed that, whether or not some deaths were inevitable, most were not and it was to these cases that they should address themselves.[67]

Theoretical agnosticism may have helped the cause of anesthetic professionalism, but it was of little use in the immediate context of the inquest. There, a clear cause of death was required where clarity was an acknowledged rarity. "The Coroner's Catechism," composed by Dudley Buxton at the request of a London coroner in 1893, was intended to provide some guidance for questioning at inquests, focusing on questions about the choice of the anesthetic agent, the condition of the patient, and the care taken by the anesthetist before and during administration.[68] Because the state of opinion on such questions as dosage, safety of agents, and reliability of preadministrative examination lacked any kind of solid consensus, however, there was the widest possible scope for the anesthetist to demonstrate adherence to some formula of recognized procedure.[69]

In the face of such indeterminacy in physiological research, inquests relied upon the one ostensibly stable piece of evidence at their disposal, the corpse. And yet the corpse was at best a limited witness at its own inquest. One of the few points on which all writers on anesthetics agreed upon was the fact that death from anesthesia left no distinct postmortem traces. This view was brought to prominence by John Snow in his 1858 work, *On Chloroform,* in which he reviewed the postmortem reports of the first five anesthetic inquests held in England and found "the appearances met with on dissection do not differ from those that are found in many other cases, especially of sudden death."[70] Testimony before the 1909 departmental committee showed that, fifty years on, Snow's views were still the received wisdom of the profession. F. W. Hewitt, asked if postmortems show the cause of death in cases of death under anesthesia, responded, "Not necessarily; in fact, usually not."[71] Spilsbury and Horsley both agreed that, in Spilsbury's words, the action of anesthetics leave "no reliable appearance."[72] When a committee member asked R. S. Trevor, a leading London pathologist, whether postmortem examinations in all cases of anesthetic death were desirable, he replied that they were for the reason that they could show that there had been no gross neglect on the part of the anesthetist: "The post-mortem has often shown appearances and lesions which would certainly have accounted for the death quite apart from the anaesthetic." "In other words," pressed the questioner, "it is an advantage to the anaesthetist himself to have a post-mortem examination?" "I certainly think so," Trevor answered.[73]

Dudley Buxton, while skeptical of the scientific value of such findings, noted the irony that the ritual loathed by anesthetists might well prove a most welcome ally in times of trouble. Writing in criticism of the retrospective emphasis typically placed on the condition of the patient's heart in attempts to explain a death under anesthesia, Buxton observed:

> The changes made in the organism by chloroform or ether cannot be dissected by the scalpel, tested by the test-tube, nor reckoned in the chemical balance. It may be a satisfaction to the person who gave the chloroform to be able to face the jury with the announcement that the patient's

heart was undoubtedly fatty; but the scientific anaesthetist cannot but remember how many must have been the fatty hearts of patients who passed safely enough the rubicon of chloroform. In the post-mortem rooms of any large hospital how many normal hearts do we find?[74]

Correspondence between two of the leading lights of British anesthetics in the early decades of the twentieth century put the point more strongly still. In a 1908 letter to Augustus Waller, A. G. Vernon Harcourt wrote: "I have been for about six years a member of the Chloroform Committee of the BMA, and have thus had opportunities of learning one of the secrets of the doctors, namely, that when a death happens the doctor says to the sorrowing relatives, a weakness of the heart showed itself which could not have been foreseen. But speaking one to another, they assume that overdosage was the cause of death."[75]

Intraprofessional acknowledgments of the limited explanatory value of such findings, of course, did nothing to dampen their popularity at inquests. The inquest, *because* of its "unscientific" nature, dependent as it was on establishing a legible medical cause of death, fell back upon a form of proof that was widely acknowledged among specialists to be at best partial. Morbid anatomy functioned in the context of anesthetic inquests as performative rhetoric. The postmortem grounded the mysterious and disturbing phenomenon of anesthetic death by ostensibly locating a tangible, organic cause for the death, a cause, moreover, that operated on a virtually inescapable logic of absence. The inquest thus combined a search for a conclusive cause of death with a broadly accepted discourse of proof whose structural relationship to the question at hand overdetermined the outcome.[76] Postmortem accounts of preexistent organic dysfunction at once confirmed medicine's capacity to serve as the master narrative of death causation, even when forced to take itself as the subject of inquiry, and provided a reading compatible with the needs of public inquiry. The inquest, not by design so much as by virtue of the relationship between its object of inquiry and its framework of investigation, produced verdicts that served as useful fictions for the anesthetist involved in a death. The scientific and the public could indeed coexist, however uneasy the terms of coexistence might be.

EPILOGUE

Legislation, though often serving historians as a privileged sign that some resolution of a prior tension has been achieved, is as likely as not to replicate within itself the very indeterminacy of the situation into which it is called to intervene. The 1926 Coroners (Amendment) Act is a case in point. In one sense this measure can be taken as the culmination of a century of medicalizing reform, composed of provisions that cast medicine, the body, and the public in new relationships to one another. Yet by itself such a reading of the act would be overly simplistic. The controversies that held the inquest in sway for so long were tenacious precisely because they were about much more than the inquest itself, and to expect an act of Parliament to resolve them would be to miss the point of what has been elaborated in the preceding pages. Thus, while outlining a new set of operational mechanisms within which coroners might forge a more efficiently bureaucratic system of death inquiry, the act at the same time absorbed into itself the vexed issue of publicity, writing into the law a tenuous balance between the needs of scientific insularity and those of public accessibility. The government's attempt to manage the claims of science and of the public, in other words, reinscribed in important respects their deeply entrenched and complex relation even as it was ostensibly being resolved.

I

The most significant way that the Coroners (Amendment) Act addressed the reformist agenda was in its inauguration of a new regime of the body at inquests. Though leaving the coroner's duty to view the body unchanged, the jury's view was made discretionary, to be carried out only if the coroner or a majority of jurors insisted. Jurisdictional constraints centered on the body were also greatly eased. The body could be removed from the immediate vicinity in which it lay to a place either within or even outside the coroner's district without prejudice to his powers or duties. The act also made the first statutory provisions for holding an inquest in the absence of its central referent; a coroner could, subject to Home Office approval, proceed with an inquest even when, "owing to the destruction of the body by fire or otherwise or to the fact that the body is lying in a place from which it cannot be recovered," there was no body to "sit upon." Finally, and most important of all, the act empowered coroners to commission postmortem examinations before deciding upon the necessity of a full (i.e., a public) inquiry.

These provisions had broad implications. Without the requirement of the jury view and without the constraints on the body's mobility imposed by the local fixity of inquest proceedings, the era of makeshift courts and postmortem rooms might be brought nearer to a close, with the dictates of efficient and accurate interrogation of the body taking precedence over both its public display and its connection to place. By uncoupling the autopsy from the inquest, moreover, the law realized the hopes of those looking for a more streamlined inquiry, one that disrupted the distorting relationship between medical and public inquiry. Medicine would get its chance to account for death before turning the question over to the confusions of public discussion.

The act also promised to delineate clearly the medical and legal components of inquests, seemingly in medicine's favor. Coroners were to adjourn any inquest case when another legal body had charged a suspect in connection with the death under inquiry. The coroner might, upon completion of these criminal proceedings, reconvene the inquest but could not record a verdict that "contain[ed] any finding which is inconsistent with the determination of any matter by the result of those proceedings."[1] The hallmark of the popular and participatory inquest and the element that lent inquest proceedings their most obvious legal form—the lay jury—was to a large extent marginalized as well, as coroners were granted the power to hold most of their inquests without a jury.

These provisions had a significant effect on the conduct of inquests. By the end of the next decade, the investigative emphasis of the system had shifted from the courtroom to the preliminary medical examination. In 1926, 26.5 percent of cases reported to coroners were dismissed with only a preliminary investigation. In 1938, by contrast, a full 46 percent of reported deaths were handled without recourse to an inquest. Of these, 42.5 percent were decided on the basis of postmortem examination.[2] When full inquests were held, they typically proceeded without a jury, and according to a Home Office survey conducted some years after the passage of the act, no jury had been known to exercise its option to view the body.[3] In a real sense, the system of inquiring into unexplained death was medicalized.

Not entirely, however. The act, in the first place, failed to deliver on several key reform demands. On the longstanding issue of the appropriate qualifications for holding the office of coroner, it made either medical or legal credentials acceptable. Fees for medical witnesses were left at the level set nearly a century earlier, and medical certificates of death remained unremunerated. Though the ability to perform postmortems without inquests increased the frequency of such examinations, general practitioners were still recognized as competent to conduct them, and the need to improve the conditions in which they were held was not addressed.[4] Furthermore, even though its provisions for preliminary postmortems and nonjury inquests were two crucial medicalizing innovations, both were accompanied by an extensive list of qualifications constraining the coroner's ability to resort to these measures. In framing these exceptions, the act appeared to leave intact the proposition that publicity was the inquest's underlying principle, with the public's involvement in the apparatus of death inquiry to be limited only in cases construed as publicly "uninteresting."[5]

More important, however, than detailing such practical limitations—which, after all, are indicative less of a fundamental tension in the logic of reform than its partial implementation—are the conclusions to be drawn from an analysis of the prolonged process through which the act was fashioned. It is here that the degree to which medicalization was a problematic resolution becomes apparent. The 1926 act, as the reader will by now be well aware, concluded several decades of legislative attention to coroners and inquests. The reports of the 1893 and 1909 parliamentary committees of inquiry were well publicized, and the arguments for their implementation were well rehearsed. But though there was a strong sense of expectation that legislative action was on the horizon, the exact shape of the new inquest system was far from clear. Parliamentary reports had tended to subsume a few striking proposals (preliminary autopsies and public certifiers, for example) under a broader framework stressing coroners' discretion. This approach faithfully reflected the Home Office's own ambivalence on the question of reform, an ambivalence derived from its intimate familiarity with the basic problem discussed throughout this book—that the inquest resided between public and scientific imperatives of inquiry that could not be fully reconciled.

Negotiations over the framing of a government-sponsored inquest bill began in earnest amid the reforming fervor of the immediate post–World War I era. The inquest had been provisionally transformed in two important respects under the wartime administration. In the name of national security, inquests in cases of death arising out of war-related domestic incidents had been significantly restricted.[6] In addition, the Jury (Emergency Wartime Provisions) Act of 1918 had empowered coroners to dispense with juries at their personal discretion. This latter innovation was made ostensibly in the interest of maximizing manpower for the war effort, though Home Office notes indicate that officials anticipated that this might result in a permanent change in inquest practice. "If this power were given to Coroners *for the duration of the war as a measure of economy*," Herbert Simpson wrote in a 1917 memorandum, "I have little doubt that it would be retained after the war for its own sake."[7] As the war drew to a close, Simpson went further still, counting the inquest jury among its casualties: "One effect of the war must have been to restrict greatly the space devoted by the Press to reports of inquests, and the absence of objection—so far as we know—to the practice adopted by many Coroners of dispensing with inquests altogether in the case of aeroplane accidents, deaths from operations of war, etc., seems to show that this practice may be properly extended to other cases."[8] Still, Simpson and his colleagues looked upon the elimination of inquest juries with some trepidation. "Where there has been much unpleasant rumour about a death or there is conflicting evidence as to its cause or personal interests will be affected by the result," Simpson wrote in an accompanying memorandum, "there should, no doubt, be a formal inquiry, but in such a case there ought to be a Jury and a verdict on oath, not merely by the Coroner."[9]

Home Office equivocation on this subject was by no means new. From at least the turn of the century, it had seen the future of the inquest jury as a key factor in its attempt to negotiate the twin ideals of transparency and scientific accuracy, for the most part deferring to the "almost superstitious regard" in which the jury was held.[10] In the wake of the 1909 parliamentary committee, the Home Office outlined an alternative to the outright elimination of the jury, one that it hoped might retain—and even enhance—the inquest's public functions while providing for a savings of the public's time, expense, and feelings. As inquest juries were most valuable "where it is desirable to satisfy public opinion," suggested the Home Office, "for this purpose the more jurors there are to agree in a verdict the more effect the verdict is likely to have. It might perhaps be possible to empower the coroner to vary the number of the jury, according to the nature of the inquiry at hand."[11] As with the subject of inquest reform writ large, officials approached the coroner's jury with care, clearly seeking to preserve what they considered to be the essence of inquest practice, some sign of public participation and public assent. But these attributes did not need to be regarded as categorical absolutes to be uniformly achieved. The imperative of transparency was flexible, subject to essentially contingent requirements that could only be gauged with sensitivity to circumstance.

These were the considerations framing the series of postwar interdepartmental meetings on inquest reform. In its first position paper, the Home Office stated its view of the problem in a suitably circumspect manner: "The question of publicity has an important bearing on the practicability of making any change in the law relating to certification of death, and Coroners' Inquests. Broadly, the public insist on publicity in all cases of doubt and suspicion, and resent it when unnecessary."[12] The complexities suppressed in such a circumlocutory formulation returned with a vengeance in a subsequent report cataloging Home Office objectives: points A to C proposed to leave common and statutory law untouched in several classes of inquests, to revert to the jury system in all inquest cases, and to regularize procedures for making inquiries before deciding on holding an inquest. The list then seems to implode under the pressure of an overdetermined notion of publicity:

> (D) to enforce a proper amount of publicity concerning action taken, or proposed to be taken, under headings (A) and (C). Publicity is also involved under heading (B).
>
> 7. As regards para. 6(D), all the speakers had in mind cases in which they thought publicity was essential.
>
> 8. Publicity, however, is not always desirable. . . .
>
> 9. If Coroners are to be encouraged, as they should be encouraged, to dispose of certain cases without holding inquests, it may be right to require publicity concerning the decision.[13]

In their attempts to balance the requirements of accurate and efficient inquests with those of a fluid notion of publicity, Home Office officials saw others' desire for reform along unambiguously medical lines as a major impediment. The Ministry of Health and its advisers in the British Medical Association were deemed the worst offenders in this regard. Arthur Locke reported after one meeting that they "spent much time in propositions for elaborate and expensive systems of 'Public Certifiers' to investigate deaths, etc, [which] would have been made a vehicle for destroying the whole system of Coroners' inquests." Had the Home Office not resisted, he concluded, health officials might have succeeded in what must have been their ultimate aim, to "dissect the inquest system."[14]

The Home Office, as a matter of principle, was committed to rejecting any simple model for resolving the inquest's longstanding tensions. It comes as no great surprise that the 1926 act's indeterminate approach to the scientific-public polarity laid the ground for continued concern. In the short term and most directly, several coroners made a point of refusing to avail themselves of the efficiency-oriented innovations. As befitting the complex map of positions adopted in the debate, it was a London-based medical coroner, F. J. Waldo, who led the charge against medicalization, lodging a protest with the Home Office at the time of the act's passage through Parliament on the grounds that it "savoured of the Scotch secret system of inquiry." In published letters and reports and in private correspondence, Dr. Waldo proclaimed his allegiance to juries, views, and full public inquiry, asserting in his 1928 Annual Report to the Corporation of London that he never dispensed with an inquest by means of a preliminary postmortem and that every one of his inquests was jury-based. "I have always acted on the principle," Waldo wrote, "that if a post-mortem examination be necessary a public inquest on oath is equally necessary if the real interests of the public are to be served."[15]

If the rear-guard action of coroners like Waldo called attention to the simple truth that much of the act's medically oriented provisions had been purposely left in the realm of the discretionary, the zeal with which other coroners rushed to embrace the inquest's new efficiencies caused the framers of the legislation far greater consternation. An avoidance of immoderate enthusiasm, after all, had been what the endless preparatory negotiations had been about. Officials were alarmed, for example, at the outcry in the press over reports that coroners were holding inquests without juries in closed courts, at times even in their own homes. Even before 1926, complaints had been made about such practices, provoking Herbert Simpson to observe: "Considerable value is attached to publicity . . . An inquest held at a Coroner's own home without notice having been given to anyone except the witnesses whose evidence he desires to have would have little value for the purpose for which inquests are at present most valued."[16] An increase in the volume of complaints received by the Home Office—accompanied by press cuttings with headlines such as "The Coroner a Law Unto Himself. Extraordinary Powers of Exclusion"—led Arthur Locke to observe that, if "the sudden tendency to hold inquests in private went on, and there was anything to suggest that it was open to criticism, the Secretary of State would have to consider whether rules of procedure should be framed. All concerned should act so as to avoid the necessity."[17] Future interventions in the conduct of inquests, Locke implied, would not necessarily take medical expediency as their starting premise. Public interest retained its place as a potent framing discourse.

The last comprehensive parliamentary inquiry into the workings of the English inquest was embodied in the 1971 Broderick commission report. In an introductory section surveying the history of the inquest, its authors pointed with satisfaction to the increasingly influential role played by medical science at inquests over the past century and a half. But despite its obvious enthusiasm for the "modern" institution that the inquest had become, the Broderick report is littered with passages that reinscribe the terms of the "traditional." Commenting, for example, on the calls made to clarify the rules for notifying coroners of potentially inquirable deaths, the committee observed:

> We live in a society in which the deaths of individuals rarely pass unnoticed or without comment. Our instinctive and continuing need to our own and other people's physical well-being is a constant stimulus to curiosity about the causes and circumstances of death. That curiosity is in itself a safeguard against the concealment of abuses, for when the facts are uncertain or suspicions are aroused, it is quickly transformed by a sense of moral duty or professional ethics into the first steps toward appropriate enquiry . . . When society acts instinctively in this way it would be as absurd as it is unnecessary to suggest replacing the present untidy set of provisions by elaborate rules.[18]

Instinct, curiosity, suspicion—the very "untidy" trio of terms that have so freely circulated in the texts considered in the foregoing pages—have perhaps not been so clearly excised from the modern medicopolitical lexicon as some might wish and others might expect.

Notes

Introduction

1. The only existing historical monograph on the modern inquest is J. D. J. Havard's policy-minded *Detection of Secret Homicide: A Study of the Medico-legal System of Investigation of Sudden and Unexplained Deaths* (London, 1960). Thomas R. Forbes, "Crowner's Quest," *Transactions of the American Philosophical Society* 68 (1978): 5–52, drawn from his review of inquest documents from the City of London and Southwark between 1788 and 1829, is informative, esp. on the categories of death subjected to inquiry. The best account of the nineteenth-century inquest, however, is to be found in a work not explicitly focused on the inquest itself—Olive Anderson's superb *Suicide in Victorian and Edwardian England* (Oxford, 1987). Lindsay Prior's *The Social Organization of Death* (London, 1989) presents a theoretically sophisticated and historically sensitive sociological account of the twentieth-century inquest in Northern Ireland.

2. As will be seen in ch. 2, debates about whether inquests were held unnecessarily were staples of the nineteenth-century discourse on inquests. The 1887 Coroners Act promised to clarify the situation, declaring the need for an inquest when a person has died "a violent or an unnatural death, or has died a sudden death of which the cause is unknown, or that such person has died in prison, or in such place or under such circumstances as to require an inquest in pursuance of any Act" (50 & 51 Vict., c. 71, s. 3 [1]). This omnibus codification of seven centuries of inquest law still left much to the coroner's discretion, providing, for example, no statutory definition of unnatural death.

3. A table showing the rate of inquests as a proportion of total deaths in England and Wales from 1856 to 1907 indicates relative stability at 5.0% to 5.5% up to the late 1880s, increasing to between 6% and 7% from 1890 to 1907. *British Parliamentary Papers* (hereafter *BPP*), 1909, vol. 15, "First Report of the Departmental Committee Appointed to Inquire into the Law Relating to Coroners and Coroners' Inquests: Pt. 2, Evidence and Appendices," app. 2, 610. The registrar general's first annual return (published in 1839) suggests a rate of 5% for 1837.

4. After the 1888 Local Government Act (51 & 52 Vict., c. 41), coroners were appointed by borough or county councils. There were other minor categories of coroners, including franchise coroners, whose appointment was in the gift of a specific individual or institution, and coroners who were responsible for highly circumscribed jurisdictions like the royal household.

5. At midcentury, e.g., 91 of the 324 coroners for England and Wales held fewer than ten inquests per year. *Lancet* 2 (16 Oct. 1858): 403.

6. Precise figures on medically trained coroners for the earlier part of this study are hard to come by, though it can be safely inferred that until the final quarter of the century their numbers were low. In 1830, the *Lancet* could not name a single active medical coroner; nearly a decade later, it counted twenty-five. From about 1880 onward, the *Medical Directory* published periodic lists of coroners, giving their occupations. According to these lists, between 1880 and 1905 about fifty (roughly 15%) of the coroners for England and Wales were medically trained.

7. In settling on a cause, the inquest might name an individual responsible for the death, in which case the finding could serve as the pretext for subsequent criminal or civil proceedings. In such cases, however, the verdict had no evidentiary standing. Although in this limited sense inquests could have criminal or civil applications, the overwhelming number of inquests resulted in verdicts without any named culpable individual; natural or accidental death was by far the commonest verdict returned.

8. In the first half of the nineteenth century, the appropriate source of information was a matter of dispute between those who regarded unofficial channels of information as an essential feature of the inquest's popular and decentralized mandate and those favoring a system more closely aligned with the expanding mechanisms of inspection and policing.

9. A number of statutory limitations were placed on the preliminary inquiry. The most significant of these prohibited the coroner from commissioning a medical examination of the corpse in question unless he had already determined to hold a full (i.e., jury-centered) inquest.

10. For contemporary observations, see, e.g., the testimony delivered before the 1860 Parliamentary Select Committee on the Office of the Coroner, *BPP*, 1860, vol. 22, questions 223, 478, 765–67, 962, and 1144. In their survey of inquests in Westminster, Maria White Greenwald and Gary I. Greenwald found that, with rare exception, neither professionals and gentlemen nor servants and unskilled laborers served as jurors. "Coroners' Inquests. A Source of Vital Statistics: Westminster, 1761–1866," *Journal of Legal Medicine* 4, no. 1 (1983): 51–86, 55. For a general discussion of the history of the coroner's jury, see W. R. Cornish, *The Jury* (London, 1968), 245–48. The 1919 Sex Disqualification (Removal) Act in principle opened up inquests to women jurors, though for the period covered in this study coroners tended to act on the advice given by the Home Office that the act did not require any change in the way that jurors were summoned. *Annual Report of the Coroners' Society*, 1920–21, 72.

11. *Cobbett's Political Register* 80, no. 9 (1 June 1833): 558.

12. Depictions of the inquest as "popular" will feature prominently in my analysis, and therefore it might be useful to lay some historiographical and conceptual foundations for the term itself. As John Belchem has shown, arguing from traditions of "popular" liberty was the cornerstone of a distinct form of English radical politics, the core elements of which included the importance of a historical sensibility animated by a prelapsarian critique of modern "corruption" and an appeal to an inclusive and participatory sense of a sovereign "people" over and above centralized and statist conceptions of governance. Against previous interpretations that had tended to dismiss such arguments from popular historical tradition as "irrational" or "primitive" precursors to a more mature, realistic consciousness of class forged in the uncompromising crucible of industrialization, Belchem's important intervention insisted on the creative and dynamic nature of the style of the politics of popular constitutionalism in the opening decades of the nineteenth century. John Belchem, "Republicanism, Popular Constitutionalism, and the Radical Platform in Early Nineteenth-Century England," *Social History* 6, no. 1 (1981): 1–32.

Historians have since extended—conceptually and chronologically—their appreciation of this powerful strand of English political culture. Gareth Stedman Jones in his seminal "rethinking" of the languages of class takes the viability of popular political vision up to the demise of Chartism, James Vernon to at least the Second Reform Act, Patrick Joyce to the turn of the century, and others to the outbreak of the First World War. For a recent overview and assessment of the historiography of popular politics, see Rowan McWilliam, *Popular Politics in Nineteenth Century England* (London, 1998), ch. 5.

13. Edward Herford gives examples of more than fifty riders issued by his Manchester juries in *The Office of the Coroner* (Manchester, 1877), app. A, 25–29.

14. For a compelling discussion of the origins and development of the theory and practice of jury mitigation, see Thomas A. Green's landmark *Verdict According to Conscience: Perspectives on the English Criminal Trial Jury, 1200–1800* (Chicago, 1985). Michael MacDonald's work on early modern suicide also emphasizes the inquest jurors' intermediary role between legal and popular attitudes. His most recent discussion is in MacDonald and Terrence Murphy, *Sleepless Souls: Suicide in Early Modern England* (Oxford, 1990), ch. 4. For a brief consideration of inquest mitigation in the context of nineteenth-century workplace deaths, see Elisabeth Cawthon, "New Life for the Deodand: Coroners' Inquests and Occupational

Deaths in England, 1830–46," *American Journal of Legal History* 33 (1989): 137–47.

15. The evidence presented to the commissioners on criminal law reform in 1845 contains a sampling of derisory descriptions of inquest juries. Examples of contemporaneous paeans to the inquest jury are given in ch. 1. The historical literature on the social composition of different types of civil and criminal juries is vast and often contentious. For an introduction to the different approaches taken and conclusions reached, see James S. Cockburn and Thomas A. Green, *Twelve Men Good and True* (Princeton, 1988), esp. the essays of John Beattie, Douglas Hay, Peter King, and Green. Beattie's *Crime and the Courts in England, 1660–1800* (Princeton, 1986), chs. 7–8, provides a comprehensive discussion of the early modern period. For a quick insight into the scholarly and political stakes of literature, see E. P. Thompson, "Subduing the Jury," *London Review of Books* 8, no. 21 (4 Dec. 1986): 7–9.

16. The question of whether inquests were open was at times a matter of fierce controversy, as will be seen in ch. 1.

17. The seventeenth-century jurist Matthew Hale had provided the classic formulation of this precept when he wrote that "the coroner's court is to inquire truly *quomodo ad mortem devenit,* and is rather for information of the truth of the fact as near as the jury can assert it, and not for an accusation." Matthew Hale, *The History of the Pleas of the Crown* (London, 1800), 2:60.

18. Sir John Jervis, *A Practical Treatise on the Office and Duties of Coroners* (London, 1829), 217. As an exception to the developing rule against admitting hearsay, for instance, witness testimony taken before the coroner could be held to be "evidence" for the purposes of further proceedings even though the individual to be placed on trial might not have been in attendance at the inquest and thus would have had no opportunity for cross-examination. As a result of this important distinction, the inquest remained largely unaffected by the formalization of evidentiary procedure attendant upon the cross-examination—most obviously the introduction of lawyers—which in other tribunals tended to mediate the "open" relationship of jurors to witnesses.

19. Until the 1926 Coroners (Amendment) Act (16 & 17 Geo. 5, c. 59), bodies could not even be transported across the subdistricts in a coroner's jurisdiction, frustrating attempts to establish central mortuary facilities and coroners' courts. For more on this, see ch. 3 and the epilogue.

20. "The Use of Inquests," *Spectator* 56 (20 Jan. 1883): 79.

21. Take, for example, the use made of Sir Thomas Smith's sixteenth-

century legal classic *De Republica Anglorum* by the leading nineteenth-century authority on inquest law, Sir John Jervis. In his brief consideration of the inquest, Smith had observed that "the impaneling of the Coroner's inquires, and the view of the body, is commonly in the street, in an open place, and *in corona populi*" (Sir Thomas Smith, "History of the Commonwealth," 96, cited in Jervis, *A Practical Treatise*, 214). Jervis cites this passage not only to indicate that the sixteenth-century inquest was a public affair, but further to suggest that Smith's depiction was directly accessible and applicable to a modern consideration of the issue. Jervis thus effects a kind of conceptual and spatial conflation, claiming that the physical "openness" of Smith's outdoor inquest might serve to elucidate questions about the inquest's relationship to a nineteenth-century conception of the "public."

For a challenging analysis of the rhetoric operations of legal arguments from "ancientness," see Peter Goodrich and Yifat Hachamovitch, "Time out of Mind: An Introduction to the Semiotics of Common Law," in Peter Fitzpatrick, ed., *Dangerous Supplements: Resistance and Renewal in Jurisprudence* (London, 1991), 159–81.

22. On expertise as "disembedding," see the work of Anthony Giddens, esp. *Modernity and Self-Identity: Self and Society in the Late Modern Age* (Cambridge, 1991), chs. 1 and 5; on the place of expert authority within the cultural negotiation of trust, see Steven Shapin's epilogue to *A Social History of Truth: Civility and Science in Seventeenth-Century England* (Chicago, 1994); on expertise as a form of the essentially modern epistemology of "abstraction" by which complex social and political realities are construed (or "reified," "commodified," and "fetishized") as knowable and manageable, see Mary Poovey, *Making a Social Body: British Cultural Formation, 1830–1864* (Chicago, 1995), esp. chs. 1, 2, and 4.

23. The term *revolution in government* is from Oliver MacDonagh's influential and controversial article, "The Nineteenth Century Revolution in Government: A Reappraisal," *Historical Journal* 1, no. 1 (1958): 52–67. For a lucid review of the MacDonagh thesis and the intense debate it generated, see Roy MacLeod's introduction to his edited volume, *Government and Expertise: Specialists, Administrators, and Professionals, 1860–1919* (Cambridge, 1988), 1–27.

24. John Stuart Mill, *Considerations on Representative Government* (Indianapolis, 1958), 91.

25. Jürgen Habermas, *The Structural Transformation of the Public Sphere: An Inquiry into a Category of Bourgeois Society*, trans. Thomas Burger (Cambridge, 1989), 236. Habermas discusses the connection be-

tween science and publicity more explicitly in "The Scientization of Politics and Public Opinion," in *Toward a Rational Society: Student Protest, Science, and Politics*, trans. Jeremy J. Shapiro (London, 1971). Recent work concerned with the "public understanding of science" has drawn on this distinction between active and passive forms of public scrutiny of scientific work. See esp. Harry Collins's distinction between "epidictic" and "experimental" forms of display in "Public Experiments and Displays of Virtuosity: The Core-Set Revisited," *Social Studies of Science* 18, no. 4 (1988): 725–48, 727–30.

In the historiography of modern England, moreover, James Vernon's innovative rereading of nineteenth-century English political culture has suggested that developments like the introduction of the secret ballot, the repeal of the stamp tax, and the reform of "corrupt" electoral practices—often taken as signal instances of a progressive and modernizing politics of inclusion—represented in important respects the gradual and uneven closure of a collective and open "public" political culture. Vernon, *Politics and the People: A Study in English Political Culture, c. 1815–1867* (Cambridge, 1995). Vernon's self-consciously bold attempt to "break the interpretive log-jam" of nineteenth-century English political history has drawn a good deal of critical attention. See, e.g., Jon Lawrence, *Speaking for the People: Party, Language, and Popular Politics in England, 1867–1914* (Cambridge, 1998), ch. 3.

26. M. Jeanne Peterson, *The Medical Profession in Mid-Victorian London* (Berkeley, 1978), 286. As Peterson herself recognizes, professionalization was by no means a straightforward process, even within the confines of establishment medicine. She (and others) have argued that the medical elite in England were loathe to supplant public codes of judgment with a narrow (but ultimately empowering) professional and scientific ethos because the former code afforded them a workable system of differentiation from the common practitioner. For other discussions of medical professionalization in the English context, see Christopher J. Lawrence, *Medicine in the Making of Modern Britain* (London, 1994); Rosemary Stevens, *Medical Practice in Modern England: The Impact of Specialization and State Medicine* (New Haven, 1966); and Ivan Waddington, *The Medical Profession in the Industrial Revolution* (Dublin, 1984). For a pan-European discussion, see W. F. Bynum, *Science and the Practice of Medicine in the Nineteenth Century* (Cambridge, 1994).

27. Ivan Illich, *Limits to Medicine. Medical Nemesis: The Expropriation of Health* (London, 1977), 12.

28. Geoffrey Gorer, *Death, Grief, and Mourning* (London, 1965);

Philippe Ariès, *The Hour of Our Death,* trans. Helen Weaver (London, 1987). Prior's introduction to his *Social Organization of Death* provides an incisive overview of the historical and sociological literature on the social contexts of death.

29. Death "becomes improper, like the biological acts of man, the secretions of the human body. It is no longer acceptable for strangers to come into a room that smells of urine, sweat, and gangrene, except to a few intimates capable of overcoming their disgust, or to those indispensable persons who provide certain services. A new image of death is forming: the ugly and hidden death, hidden because it is ugly and dirty." Ariès, *The Hour,* 569.

30. This genealogy of medical perception, which figures prominently in works as analytically diverse as Erwin Ackerknecht's *Medicine at the Paris Hospital, 1794–1848* (Baltimore, 1967) and Michel Foucault's *The Birth of the Clinic: An Archeology of Medical Perception,* trans. A. M. Sheridan Smith (New York, 1973), runs from Morgagni's localization of disease and death in organs, to Bichat's in tissue structure, to Virchow's in cells.

31. Foucault, *Birth of the Clinic,* 35. The classic discussion of "normality" as a constitutive feature of modern medicine is Georges Canguilhem, *On the Normal and the Pathological,* trans. Carolyn R. Fawcett (New York, 1989). For a more recent consideration of the workings of the norm in the social history of medical practice, see John Harley Warner, *The Therapeutic Perspective: Medical Practice, Knowledge, and Identity in America, 1820–1885* (Cambridge, Mass., 1986).

32. This approach has clear parallels with recent studies in the sociology of scientific knowledge that have emphasized a strategic and negotiated relation between scientific activity and its public. Of these studies, the work of Simon Schaffer and Steven Shapin on the culture of seventeenth-century experimental science resonates most strongly with the terms of my analysis. They consider the legitimation of scientific practice to be fundamentally constituted in relation to a notion of "the public." For a quick introduction to the terms of this analysis, see Steven Shapin, "Science and the Public," in R. C. Olby, G. N. Cantor, J. R. R. Christie, and M. J. S. Hodge, eds., *Companion to the History of Modern Science* (London, 1990), 990–1007. For more detailed examples, see Shapin, "Pump and Circumstances: Robert Boyle's Literary Technology," *Social Studies of Science* 14 (1984): 481–520; Shapin, "'A Scholar and a Gentleman': The Problematic Identity of the Scientific Practitioner in Early Modern England," *History of Science* 24 (1991): 279–327; and Shapin and Simon Schaffer, *Leviathan and the Air-Pump* (Princeton, 1985).

Chapter 1: The Genealogy of the Popular Inquest

1. *The Times,* 14 September 1830, 4.

2. *Morning Chronicle,* 10 September 1830, 3.

3. *The Times,* 10 September 1830, 4.

4. *Lancet* 2 (2 Oct. 1830): 43.

5. *Morning Chronicle,* 10 September 1830, 3.

6. *The Times,* 14 September 1830, 4.

7. Wakley laid several charges of interestedness: against lawyers in general (that they had a professional interest in verdicts of manslaughter, for example) and Baker in particular (dismissing him as a mere creature of vestry politics). *Morning Chronicle,* 21 September 1830, 3; 18 September 1830, 4.

8. Ibid., 15 September 1830, 1.

9. *The Times,* 13 September 1830, 4. George Frederick Young was a London shipping merchant who entered the House of Commons in 1832.

10. Ibid.

11. Important recent work on the prison inquest has been undertaken by historians and social analysts seeking to provide perspective on several controversial deaths in custody dating from the early 1980s. Much of this work has been done in conjunction with the policy research group Inquest, set up in 1981 at a grassroots level to investigate, monitor, and report on procedures for inquiring into politically sensitive deaths. The explicitly present-day orientation of Inquest and its allied scholars has tended to push history into the background, with certain exceptions: Tony Ward has written informative articles on the subject of nineteenth-century prison inquests, including "Coroners, Police, and Deaths in Custody in England: A Historical Perspective," *Working Papers in European Criminology* 6 (1985): 186–215; and with Joe Sim, "The Magistrate of the Poor? Coroners and Deaths in Custody in Nineteenth-Century England," in Michael Clark and Catherine Crawford, eds., *Legal Medicine in History* (Cambridge, 1994), 245–67. Joe Sim's *Medical Power in Prisons: The Prison Medical Service in England, 1744–1988* (Milton Keynes, England, 1990) is a thorough history of the increasing scope for medical authority in prisons and pays close attention to issues of abuse and death in custody. For an overview of the research and conclusions of the Inquest group, see Phil Scraton and Kathryn Chadwick, *In the Arms of the Law: Coroners' Inquests and Deaths in Custody* (London, 1987).

12. *Lancet* 2 (28 Aug. 1830): 873–74. Hunt was imprisoned at Ilchester Gaol for leading the 1819 mass meeting at Manchester's St. Peter's Square,

which was violently suppressed by the authorities. For an account of Hunt's place in the annals of early nineteenth-century political radicalism, see John Belchem, *"Orator" Hunt: Henry Hunt and English Working Class Radicalism* (Oxford, 1985).

13. John le Breton, *Britton*, ed. F. M. Nicholas (Oxford, 1865), 1:45.

14. Useful histories of the origins and development of English prisons in the medieval and early modern periods include Ralph B. Pugh, *Imprisonment in Medieval England* (Cambridge, 1968); Christopher Harding, Bill Hines, Richard Ireland, and Philip Rawlings, *Imprisonment in England and Wales: A Concise History* (London, 1985); and Wayne Joseph Sheehan, "The London Prison System, 1666–1795," PhD diss., University of Maryland, 1975. Sean McConville's epic two-volume history comprehensively covers penal policy, administration, and practice from the early modern period to the end of the nineteenth century. McConville, *A History of English Prison Administration* (London, 1981) and *English Local Prisons, 1860–1900: Next Only to Death* (London, 1995). Michael Ignatieff's *A Just Measure of Pain: The Penitentiary in the Industrial Revolution, 1750–1850* (New York, 1978) is a more conceptual treatment of the English penal system in the modern period.

15. John Lilburne, "London's Liberty in Chains Discovered" (London, 1646), 6–7; James Whiston, "England's Calumnies Discover'd" (London, 1696), 16; John Howard, *The State of the Prisons in England and Wales* (Warrington, 1777), 66–68.

16. Laurence Braddon, "Murther Will Out" (London, 1692); Robert Ferguson, "An Enquiry into, and Detection of the Barbarous Murther of the Late Earl of Essex, or A Vindication of that Noble Person from the Guilt and Infamy of having destroy'd Himself" (London, 1684).

17. Jacob Ilive, *Reasons offered for the reformation of the House of Correction in Clerkenwell* (London, 1757), 4.

18. Thomas Ellwood, *The History of the life of Thomas Ellwood*, 2d ed. (London, 1714), 164–65. Emphasis original.

19. In making this observation I am not insisting that there were no real examples of "popular" political inquests before the nineteenth century. There are indications that, at least at a theoretical level, this notion existed previously. The distinguished early eighteenth-century jurist William Hawkins, for example, suggested that coroners "may and ought to inquire of the death of all persons whatsoever who die in prison, to the end that the public may be satisfied, whether such persons came to their end by the common course of nature, or by some unlawful violence, or unreasonable hardships put on them by those under whose power they were confined." Hawkins,

A Treatise of the Pleas of the Crown (London, 1721), 2:47. What is certain, however, is that the popular inquest of nineteenth-century radicals drew on an ancient pedigree without ancient example, that the founding images of the profoundly historically conscious discourse of popular radicalism were, in other words, strictly contemporary.

20. Frederick Pollock and Frederic W. Maitland, *History of English Law* (Cambridge, 1898), 1:534. Since fines were the most common form by which redress was exacted, the coroner has been seen by many recent legal historians as first and foremost a Crown revenue collector. S. F. C. Milsom, for example, describes the coroner as essentially a "permanent local accountant," but adds that, since "law and order on the national scale were first expressed in terms of revenue," this description by no means diminishes the coroner's significance. Milsom, *The Historical Foundations of the Common Law* (London, 1981), 28. For a learned overview of the functions of the early coronership, see R. F. Hunnisett, *The Medieval Coroner* (Cambridge, 1961).

21. Owners of objects—cartwheels, knives, and swords, for example— which "moved to the death" of an individual were subject to a *deodand,* a fine reflecting the value of the object or objects in question. An important part of the inquest's duty in cases such as these was to assess the value of the implicated objects, and this fact accounts for the attention to a death's contextual circumstances dictated by medieval and early modern inquisition forms. On the theory and practice of deodands, see Jacob J. Finkelstein, "The Goring Ox: Some Historical Perspectives on Deodands, Forfeitures, Wrongful Death, and the Western Notion of Sovereignty," *Temple Law Quarterly* 46, no. 2 (1972–73): 169–290, and E. J. Evans's now classic study, *The Criminal Prosecution and Capital Punishment of Animals: The Lost History of Europe's Animal Trials* (London, 1987, reprinted from the 1906 ed.), ch. 2.

22. This is not to suggest that medieval coroners were simply valuers of a special segment of the Crown's revenues. Our knowledge of their role, derived almost exclusively from administrative records, tends to be biased toward the formalistic side of their duties and necessarily passes over their possible significance in other areas—in the social and political workings of local communities, for example. Work done by historians of criminal law, most importantly by Thomas A. Green, provides glimpses of this more nuanced role, suggesting that, in performing a range of ancillary duties in the name of the locality (sitting on juries and representing the community at other points of contact with the Crown's legal apparatus, for instance), coroners were positioned as intermediaries between Crown and (varied)

local interest. Green, *Verdict According to Conscience: Perspectives on the English Criminal Trial Jury, 1200–1800* (Chicago, 1985), ch. 2 passim.

23. Hunnisett's meticulous research suggests that medieval coroners rarely performed the full range of duties assigned them by law commentators, but this does not detract from the significance of discussions of the inquest's theoretical location in medieval conceptions of sovereignty. See esp. Hunnisett, "The Origins of the Office of Coroner," *Royal Historical Society Transactions,* 5th ser., 3 (1958): 85–104.

24. Henri de Bracton, *De Legibus et Consuetudinibus Anglie,* ed. Sir Travers Twiss (London, 1879), 2:269, 2:275.

25. Sir Thomas Smith, *De Republica Anglorum,* ed. Mary Dewar (Cambridge, 1982), 108.

26. John Wilkinson, *A Treatise Collected out of the Statutes of this Kingdome, and according to common experience of the Lawes, concerning the Office and Authorities of Coroners and Sherifes* (London, 1638), 6, 4.

27. "All prisons," Coke remarked in his *Second Part of the Institutes,* "are the King's" (London, 1797), 87. Even when, as increasingly became the case, prisons were franchised out to individuals, they still remained *prisona regis,* with the franchise holder both representing sovereign authority and answerable to it. By the reign of Edward I, the established doctrine was that, for a prison to be lawful, it required express sanction of the Crown: a prison, according to *Britton,* was defined by the king as "a place limited by us within certain bounds for the keeping of the bodies of men" (*Britton* 1:43).

28. John Langbein, *Torture and the Law of Proof: Europe and England in the Ancien Régime* (Chicago, 1977), 29.

29. This was clearly what Berkshire officials had in mind when they addressed themselves to the king concerning deaths due to starvation and disease in the Windsor jail during the early fourteenth century. They cast their appeal for improvement of jail conditions in terms of the king's own sovereign interest, observing that "the prisoners frequently die from want of sustenance before judgement is rendered"—that is, before fines could be collected. Prison death here represented a special (and remediable) loss of Crown revenue. Harding et al., *Imprisonment in England and Wales,* 41.

30. For an introduction to medieval and early modern conceptions of law and sovereignty, see Edward Powell, *Kingship, Law, and Society: Criminal Justice in the Reign of Henry V* (Oxford, 1989), esp. ch. 1.

31. *Hansard Parliamentary Debates* 35 (Mar. 1800–Oct. 1801): 467.

32. Ibid., 35:468.

33. Thus, when in June 1812 the Lincoln jail came under the scrutiny of

parliamentary reformers, the conduct of the county coroner was cast as part of the problem. Brougham told the House that inquest juries there "were sworn men under the influence of the governor of the prison" (*Hansard*, 23 [May–July 1812]: 312) and that "the coroner behaved throughout in a manner which was completely reprehensible, and treated the evidence in particular in a very unbecoming way" (758). Burdett demanded another parliamentary inquiry into prison administration, observing that the common thread linking abuses from Cold Bath Fields to Lincoln was the modern tendency, in an age of government repression, to prison secrecy: "There was no safety to prisoners but in the free admission of their friends to them. It was impossible to devise any other means by which they could be secured from oppression" (ibid., 760).

34. *The Times*, 14 July 1802, 3. The 1802 contest has been enshrined as a landmark of radical resistance to postrevolutionary reaction; Thompson, for one, described it as the period's "most sensational election." E. P. Thompson, *The Making of the English Working Class* (Harmondsworth, England, 1968), 493. See also J. Ann Hone, *For the Cause of Truth: Radicalism in London, 1796–1821* (Oxford, 1982), 117–33.

35. *Cobbett's Weekly Political Register* 22, no. 2 (11 July 1812): 33. Emphasis original.

36. *Cobbett's Weekly Political Register* 80, no. 10 (8 June 1833): 624; 80, no. 9 (1 June 1833): 558. Emphasis original.

37. Cobbett was not the only late convert to the timeless principles of the popular inquest. In a *Lancet* editorial a mere two years before his 1830 electoral bid, Wakley himself seemed unimpressed, dismissing suggestions to appoint medical assessors to assist legal coroners as "advising the repair of an old building, when the cheaper and wiser course would be to pull it down, and re-erect a new edifice on its site." *Lancet* 2, 26 July 1828, 534. Francis Place was another prominent radical who followed Cobbett's trajectory from ignorance to embrace. In 1815, Place was summoned as a juror in the inquest on the Duke of Cumberland's murdered valet, Joseph Sellis. According to the published recollections of an associate, Place "told him that when he received his summons from the coroner, he did not know what he ought to do, never having been summoned on such a jury before; that he went immediately to Clifford, the barrister, Sir Francis Burdett's friend, to be instructed by him in the duty and privileges of a coroner's Jury." A. Aspinall, *Correspondence of George, Prince of Wales, 1770–1812* (London, 1970), vol. 7, app. 1, p. 408. By 1830, Place was secretary of Wakley's election campaign. For details of the Sellis affair and its significance, see Iain McCalman, *Radical Underworld: Prophets, Revolutionar-*

ies, and Pornographers in London, 1795–1840 (Oxford, 1993), 34–41. As McCalman observes, many Radicals suspected the libertine and arch-conservative duke of having committed the crime and, when an open verdict was returned by the jury with Place as its foreman, they (Hunt and Burdett included) were convinced that Place had been bought off by the administration. I thank Prof. McCalman for bringing the Sellis affair to my attention and for generously sharing his ideas and research notes on the subject.

38. Dror Wahrman, "Public Opinion, Violence, and the Limits of Constitutional Politics," in James Vernon, *Re-reading the Constitution: New Narratives in the Political History of England's Long Nineteenth Century* (Cambridge, 1996), 83–122. See also John Belchem, "Republicanism, Popular Constitutionalism, and the Radical Platform in Early Nineteenth-Century England," *Social History* 6, no. 1 (1981): 1–32, and James Epstein, *Radical Expression: Political Language, Ritual, and Symbol in England, 1790–1850* (New York, 1994).

39. For more on these issues, see Burney, "Making Room at the Public Bar: Coroners' Inquests, Medical Knowledge, and the Politics of the Constitution in Early-Nineteenth-Century England," in Vernon, *Re-reading the Constitution,* 123–53.

40. The legality of publishing testimony from an ongoing inquest was itself the main issue of contention in *R. v. Fleet* (1818), at which the court decided in favor of suppression. Justice Bayley considered that such publication "may influence the public mind," rendering it impossible for jurors to "decide solely on the evidence which they hear on the trial." 1 B.& Ald. 384.

41. *Hansard* 41 (Nov.–Feb. 1819–20): 1184.

42. Letter from "Vindicator," *Examiner,* 24 October 1819, 686. "Vindicator's" position had been foreshadowed by the *Examiner's* editorial of the previous week warning that a successful suppression of inquiry would usher in an era of true Continental despotism, with "the police of this country reduced to the mixed laxity and severity old Lisbon itself—lax as far as the people's lives are to be secured, and severe only in behalf of Government."

43. *The Times,* 23 August 1821, 2. For discussions of the significance of the "Queen Caroline Affair," see Anna Clark, "Queen Caroline and the Sexual Politics of Popular Culture in London, 1820," *Representations* 31 (1990): 47–68; Thomas W. Laqueur, "The Queen Caroline Affair: Politics as Art in the Reign of George IV," *Journal of Modern History* 54 (1982): 417–66; and Wahrman, "Public Opinion," in Vernon, *Re-reading the Constitution.*

44. *Black Dwarf,* 10, no. 13 (26 Mar. 1823): 456; 10, no. 18 (30 Apr. 1823): 635–6. In July *The Times* devoted several columns to the "death of another Convict," during which it reported on the principled (if ultimately futile) resistance of jurors to the coroner's attempts to secure an uncontroversial verdict: "The Coroner impressed strenuously upon the minds of the jurymen . . . the discontent such a verdict would occasion out of doors, and ultimately succeeded in getting the whole to sign a verdict of 'Natural Death, occasioned by diarhoea,' which was recorded." *The Times,* 19 July 1823, 3. The coroner's concern was reiterated at a parliamentary committee inquiring into this death, when the jury foreman responded in the negative to a member's question, "Were you aware, in the verdict you were about to give, you would excite in the mind of the public more than common interest, as to this prison?" *British Parliamentary Papers* (hereafter *BPP*), 1823, vol. 5, "Report on the State of the General Penitentiary at Millbank," 700. Largely as a response to the Millbank controversy, the government included in its 1823 Gaol Act a provision banning the standard practice of having prisoners comprise one-half of the inquest jury. It was not until 1865, however, that the presence of prison officers on juries was outlawed.

45. *Real John Bull,* 10 August 1823, 252, and 26 August 1823, 340.

46. *The Times,* 23 October 1823, 2.

47. Henry Hunt, *Investigation at Ilchester Gaol, in the County of Somerset, into the Conduct of William Bridle, the Gaoler, Before the Commissioners Appointed by the Crown* (London, 1821). See Belchem's *"Orator" Hunt,* 134–43, for an account of Hunt's incarceration at Ilchester.

48. Hunt, *Investigation,* 167.

49. Ibid., 169.

50. When informed by prison staff that the deceased had been ironed (i.e., shackled) "in the usual way," e.g., Cames had accepted this description at face value. Hunt professed disbelief at this lack of basic investigative rigor: "And you, as a coroner, finding a man in a cell who had been ironed in the usual way, did not think it proper to ask what the usual way was?" Ibid., 168.

51. Ibid., 169.

52. Hunt questioned Bryar about the refractory prisoner Thomas Gardner, to whom Bryar had applied a head blister at Bridle's request. "Do you mean, as a medical man of character, to swear that you applied the blister at the time for a disease, or was it for a punishment?" Hunt demanded. "I certainly did in a great measure as a punishment, and that it might do him

good," Bryar replied, to which Hunt rejoined, "What do you mean by doing him good, was it his health or his manners?" "His manners," Bryar simply replied. Ibid., 61.

53. Ibid., 171. In making his demand for an independent postmortem, Hunt was well in advance of official policy; it was not until 1881 that the Home Office issued a (controversial) circular requiring such an independent investigation in all cases of prison death. See the correspondence contained in Home Office (HO) 144/265/A57408 and HO45/9589/91024 at the Public Record Office (PRO).

54. *Jurist, or Quarterly Journal of Jurisprudence and Legislation* 1 (1827–28): 46.

55. "Observations on the Nature and Importance of Medical Jurisprudence," *London Magazine* 1 (Feb. 1820): 132–36, 133, 134.

56. In Bentham's entire posthumously edited corpus, there are only a handful of references to the inquest, each of which is unequivocally disparaging. In one, he fulminated against the inquest as an inherited institution hopelessly bound by and reflective of "the inexperienced and barbarous ages in which it took its origin." *Constitutional Code,* bk. 2, ch. 26, sect. 5: "Death-Recordation Function," art. 18, in John Bowring, ed., *The Works of Jeremy Bentham* (Edinburgh, 1843), 9:629.

57. *Panopticon; or, The Inspection House,* in Bowring, *Works,* 4:46. Emphasis original.

58. Bentham was clear that there was nothing special about the form of curiosity he was marshaling in the name of rational reform: untroubled by the implicit association on which his question depended, he demanded in the *Panopticon* if it was "possible that a national penitentiary-house of this kind should be more at a loss for visitors than the *lions,* the *wax-works,* or the *tombs?*" Ibid., 4:133. Emphasis original.

59. "Of Publicity," ch. 2 of *Essay on Political Tactics,* in Bowring, *Works,* 2:310. Emphasis original.

60. Ibid., 2:310.

61. Ibid., 2:314. Emphasis original. "Whom ought we to distrust," he asked in defense of a public manifestly suspicious of its governors, "if not those to whom is committed great authority, with great temptations to abuse it?"

62. Ibid., 2:313. Bentham did not consign society to a regime of perpetual suspicion; instead, he looked to this process as a way of effecting a rationalization of the public mind and ultimately of creating a new social subject. The habit of participation in and access to the whole of the evi-

dence in question ultimately produced a citizenry capable of *withholding* judgment where appropriate: "A habit of reasoning and discussion will penetrate all classes of society. The passions, accustomed to a public struggle, will learn reciprocally to restrain themselves; they will lose that morbid sensibility, which among nations without liberty and without experience, renders them the sport of every alarm and every suspicion." Truth, then, was both induced through and stabilized within mechanisms of public access. Ibid., 2:311.

63. *Hansard,* 3rd ser., 13 (May–July 1832): 925. The debate took place on 20 June, only days after Bentham's death.

64. Ibid., 13:924.

65. Ibid., 13:926. Hunt seconded O'Connell's historical reading: "The right of excluding the public from Coroner's Inquests was first assumed, twelve years ago, at Manchester." He concluded by likening modern, *in camera* inquests to the archetypal absolutist perversion of the true constitution: "A Coroner's Inquest, as the law now stood, was little better than the Star Chamber or the Inquisition" (927).

66. Ibid., 13:937. Campbell, whose rise through the judicial ranks led him ultimately to the lord chancellorship, was in his next appointment, as solicitor general, forced to reverse his resolute defense of inquest independence in the context of the Cold Bath Fields affair.

67. The only monograph on the subject is that of the former London coroner Gavin Thurston, *The Clerkenwell Riot: The Killing of Constable Culley* (London, 1967). Because of the inquest jury's verdict, the Calthorpe Street affair has also been taken as a representative instance of a viable (though embattled) tradition of English liberties, as in E. P. Thompson's reflections on the episode in *Writing by Candlelight* (London, 1980), 205–9.

68. *Examiner,* 26 May 1833, 321.

69. *Bell's New Weekly Messenger,* 19 May 1833, 236.

70. *The Times,* 22 May 1833, 3.

71. Both reprinted in the *True Sun,* 21 May 1833, 4. The Cully inquest had a lasting legacy. In its assessment of the state of the police at midcentury, for instance, the *Edinburgh Review* recalled its troubled beginnings by reference to the injudicious actions of a politically immature jury:

The storm of clamour was so great, that it was a question whether the new system would not be borne down by the torrent of unpopularity. A striking instance of this occurred on the occasion of the Cold Bath Fields Meeting in 1833 . . . The Coroner's Inquest brought in a verdict of 'justifiable homicide.' The verdict, which ought to have been 'wilful murder,' was of course at once set aside by the Superior Courts . . . ; but the record remains to show to what an outrageous length of absurdity a Coroner's Inquest can go when it becomes the mouthpiece of a mob.

Edinburgh Review 96 (July 1852): 7.

72. *Poor Man's Guardian* 2, no. 103 (25 May 1833): 164.

73. "Instructions from the Coroner for Middlesex," reprinted in *Lancet* 1 (2 Nov. 1839): 205–6. As a leading figure in the parliamentary fight against the new Poor Law (which he described on the Commons floor as "a sanguinary and a mercenary act, based on ferocious and savage principles—an act calculated only to produce misery and torture among the deserving poor"), it is not surprising that Wakley honed in on deaths taking place in workhouses as meriting the closest scrutiny. *Hansard*, 3rd ser., 54 (Jan.–Mar. 1841): 390.

74. S. Squire Sprigge, *The Life and Times of Thomas Wakley* (London, 1889), 384. Sprigge's biography provides engaging accounts of this and other of Wakley's standout cases.

75. *BPP*, 1841, sess. 1, vol. 21, "Copies of notes and papers relating to an inquest held before Mr. Wakley upon a pauper who died in the Hendon Union Workhouse," 266.

76. *The Times*, 6 November 1840, 4.

77. *The Times*, 17 November 1840, 4. For a recent and powerful account of the diverse political coalition forged in opposition to the new Poor Law, which puts Wakley's brand of medical activism center stage, see Christopher Hamlin, *Public Health and Social Justice in the Age of Chadwick* (Cambridge, 1998), esp. chs. 2–5. Hamlin discusses workhouse deaths and the wider politics of inquiry into starvation at pp. 143–47 and in "Could You Starve to Death in England in 1839? The Chadwick-Farr Controversy and the Loss of the 'Social' in Public Health," *American Journal of Public Health* 85 (1995): 856–66.

78. Erasmus Wilson was a lecturer in anatomy and physiology at the Middlesex Hospital and a noted authority on the pathology of the skin.

79. *The Times*, 28 July 1846, 7. Wilson's detailed accounts of the case and his postmortem findings were published in the *Lancet* 2 (15 Aug. 1846): 176–77 and (14 Nov. 1846): 540–42.

80. *The Times*, 28 July 1846, 7.

81. *The Times*, 4 August 1846, 8.

82. *Daily News*, 14 August 1846. Another meeting solicited subscriptions for a plaque in honor of Wakley, whose actions at the White inquest (which were likened to his earlier stalwart defense of the "Tolpuddle Mar-

tyrs") had "proved him to be a warm friend and advocate of the poor." *Morning Chronicle,* 5 September 1846.

83. *Morning Chronicle,* 5 September 1846.

84. *The Times,* 1 August 1846, 8, 4.

85. For an important new consideration of this polarization and its underlying logic, see Dror Wahrman, *Imagining the Middle Class: The Political Representation of Class in Britain, c. 1780–1840* (Cambridge, 1995), ch. 10.

86. "Coroners' Courts," *Champion and Weekly Herald,* 2 June 1838, 110–11.

87. *Champion and Weekly Herald,* 19 March 1837, 212.

88. *Northern Star,* 26 December 1840. Emphasis original. The Chartist press had reason to be specially sensitive to the issue of institutional deaths at this time, given the wholesale incarceration of the movement's leaders after the turbulent "physical force" demonstrations of 1839–40.

89. *Journal of Social Science* 1 (1865–66): 129.

90. Henry Cartwright, "Should not Coroners be obliged by Law to hold Inquests in all Cases of Deaths within Union Poor-houses?" *Transactions of the National Association for the Promotion of Social Science,* Manchester Meeting, 1866 (London, 1867), 229. Mary Poovey has recently argued that the National Association for the Promotion of Social Science (NAPSS) functioned as an essential forum from which to act out an apolitical ideal of social policy distanced from the highly charged debates of the 1830s and 1840s. This would, of course, make it the perfect vehicle for the inquest's domestication. Poovey, *Making a Social Body: British Cultural Formation, 1830–1864* (Chicago, 1995), 97. On the workings of this organization, see also Lawrence Goldman, "A Peculiarity of the English? The Social Science Association and the Absence of Sociology in Nineteenth-Century Britain," *Past and Present* 114 (1987): 133–71.

91. Joseph J. Pope, "On the Advantages likely to accrue from a more extended recognition of the powers and work of the Coroner's Office," *Transactions of the NAPSS,* 1866, 235.

92. "The Extension of Coroners' Jurisdiction; Discussion," ibid., 291.

93. Olive Anderson, *Suicide in Victorian and Edwardian England* (Oxford, 1987), 18, provides a complete list of the relevant legislation by which lunatic asylums (1853 and 1862), prisons (1865), retreats for habitual drunkards (1879), and infant-minding institutions (1872 and 1897) were statutorily placed under the coroner's watch. Two points should be noted: first, workhouses are absent from this list of protected spaces; second, except in the already long-established procedures for prison inquiry,

these laws required only notification to the coroner of a death, not an actual inquest.

94. Reynolds has figured in several recent historical accounts of the political culture of post-Chartist radicalism. See, e.g., Rowan McWilliam, "The Mysteries of G. W. M. Reynolds: Radicalism and Melodrama in Victorian Britain," in Malcom Chase and Ian Dyck, eds., *Living and Learning: Essays in Honour of J. F. C. Harrison* (Aldershot, 1996), 182–98; Margot C. Finn, *After Chartism: Class and Nation in English Radical Politics, 1848–1874* (Cambridge, 1993), 107–18; and Miles Taylor, *The Decline of Radicalism, 1847–1860* (Oxford, 1995), 115–23. Reade has been less well served by historians but has been the subject of several literary critics. The standard work is Wayne Burns, *Charles Reade: A Study in Authorship* (New York, 1961); see also Winnifred Hughes, *The Maniac in the Cellar: Sensation Novels of the 1860s* (Princeton, 1980), esp. ch. 1.

95. Letter from "Northumbrian," *Reynolds's Newspaper,* 22 January 1865, 3. The same writer found no change when, five years on, he invoked the language of decay to describe the vacuous level to which investigations of mining accidents had descended: "Inquests have, in fact, degenerated into a very solemn farce, ending always in the discovery that some incompetent person, now dead, did something wholly unjustifiable, it is scarcely possible to say what exactly, but it is assumed to be this, or that, or the other, whatever happens to be the most convenient theory suggested at the time." "'Crowner's Quest Law' for Miners," *Reynolds's Newspaper,* 30 January 1870, 3.

96. Charles Reade, *It Is Never Too Late to Mend,* 3 vols. (London, 1856), 2:119.

97. In the coverage devoted to it in the mainstream and professional press, especially intense in the early 1870s, the issue of rib breaking in asylums was clearly taken as a troubling trend. After one case at the Hanwell Asylum, *The Times* remarked that it was "very extraordinary that patients in Lunatic Asylums should be so constantly likely to get their ribs broken," adding that "the public mind has been startled by recent disclosures" (4 Apr. 1870, 9). After some initial skepticism, the *Lancet* joined those who regarded the matter as a serious indictment of a rapidly expanding asylum system, observing in one of a series of editorials in 1870 "that fractures of the sternum and ribs are of frequent occurrence in asylums, that they must necessarily be caused by direct violence, that this violence must often be intentional, and that the system under which asylums are conducted renders it impossible to subject attendants to proper discipline or control" (*Lancet* 1 [5 Feb. 1870]: 199–200, 200). The question was frequently

broached, if rather more defensively, in the pages of the *Journal of Mental Science* during the early 1870s.

98. Charles Reade, "How Lunatics' Ribs Get Broken," *Pall Mall Gazette,* 20 January 1870, 6.

Chapter 2: Registers of Death

1. Christopher Hamlin's important study suggests that medicine's social and political radicalism in the 1830s was reflective of medicine's "traditional" concern with the individual and his or her environmental circumstances, an approach that was rendered "radical" only in relation to the marginalization of the individual under the early Victorian public health regime dominated by Edwin Chadwick. Hamlin, *Public Health and Social Justice in the Age of Chadwick* (Cambridge, 1998), esp. ch. 2. My discussion of the radical impulse in the medicine of Wakley and his contemporaries includes this traditionalist dimension, but also their more self-conscious embrace of the radical critique of "Old Corruption" in challenging the bastions of established medical authority. For a compelling depiction of this second form of medical radicalism, see Adrian Desmond, *The Politics of Evolution: Morphology, Medicine, and Reform in Radical London* (Chicago, 1989), esp. ch. 3.

2. 4 Edw. I, stat. 2 (1275–76). Hunnisett argues that this is not a statute but rather an extract from Bracton's *De Legibus,* which, because of Bracton's immense authority, quickly came to be invested with statutory standing. R. F. Hunnisett, *The Medieval Coroner* (Cambridge, 1961), 15.

3. In seeking to limit the scope of the coroner's investigative competence, the magistrates were acting out one scene in a long saga of antagonism between the county bench and coroners, an antagonism that intensified in the first half of the nineteenth century as magistrates began to use their powers of financial oversight of inquests (statutorily granted in 1751) to influence both their subject matter and their format. On the origins of the protracted struggle between these two offices, see Hunnisett, *The Medieval Coroner,* 190–99.

4. *R. v. Kent JJ.* (1809), 11 East. 229.

5. E.g., *R. v. Great Western Railway* (1842), 3 Q.B. 333; *R. v. Carmarthenshire JJ.* (1847), 10 Q.B. 796; Sir John Jervis, *A Practical Treatise on the Office and Duties of Coroners* (London, 1829), 24–25.

6. By 1860, seven English counties had passed resolutions to refuse payment for inquests returning verdicts of "natural death" or "death by the Visitation of God" in the absence of prior suspicion of criminal or other unnatural circumstances. Several others had passed on instructions to their

county constabularies not to inform coroners of deaths unless there was a suspicion of foul play. *British Parliamentary Papers* (hereafter *BPP*), 1860, vol. 57, "Orders and Regulations made, and Regulations passed, by County Magistrates in England and Wales, since 1850, relating to Expenses of holding Coroners' Inquests: —and Instructions and Directions given by Chief Constables to County Constabulary, with reference to Their Duty in giving Information of Deaths, etc.," e.g., Durham County Constabulary Orders (1857), 346.

7. Ruth Richardson, *Death, Dissection, and the Destitute* (London, 1988).

8. The Henrician statute (32 Hen. 8, c. 42) allowed six bodies per year; two additional bodies were authorized by a royal grant of Charles II. The 1752 act for "better Preventing the horrid Crime of Murder" (25 Geo. 2, c. 37) gave discretion to judges whether the body of an executed felon was to be dissected or gibbeted, which, according to Richardson, eased but by no means solved the supply problem. Richardson, *Death*, 32–37. Peter Linebaugh has demonstrated the potent and ritualized popular hostility to the place of dissection in the theater of punishment. Linebaugh, "The Tyburn Riot against the Surgeons," in Douglas Hay, Peter Linebaugh, John G. Rule, E. P. Thompson, and Cal Winslow, eds., *Albion's Fatal Tree* (New York, 1975), 65–118. Richardson further argues that it was not only the poor who felt the need to protect the body from postmortem violation; in support of this claim, she points to, among other things, the growth in demand for costly, strong vaults and patent coffins. Richardson, *Death*, 98.

9. Richardson chronicles some of the early protest against the implementation of the Anatomy Act (2 & 3 Will. 4, c. 75) and argues that this reaction lingered as an occluded part of popular memory well into the twentieth century (e.g., ch. 9, esp. 228–33, and ch. 11, 263–66). "Over the course of Victoria's reign," she observes, "the fact that the misfortune of poverty could qualify a person for dismemberment after death became too intensely painful for contemplation; became taboo. The memory went underground of a fate literally unspeakable." Richardson, *Death*, 281.

10. *Morning Herald*, 30 September 1839, reprinted in the *Lancet*, 1 (2 Nov. 1839): 206. Emphasis original. A letter to *The Times* echoed the longstanding charge of medical violation: "The loss of one's relation is distressing enough without (unless occasion should absolutely require it) an additional harrowing of the feelings by the coroner's inquest, and, perhaps, the application of the knife." *The Times*, 5 October 1839, 6.

11. For a brilliant analysis of early Victorian concerns about this form of market activity, see Catherine Gallagher, "The Bio-economics of *Our*

Mutual Friend," in Michel Feher, ed., *Fragments for a History of the Human Body* (New York, 1989), 3:344–65.

12. *Observer,* 6 October 1839; *Morning Advertiser,* 9 October 1839; *Morning Herald,* 15 October 1839; all reprinted in *Lancet* 1 (2 Nov. 1839): 205–15. Emphasis original.

13. Wakley was in no doubt as to the magistrates' motives and was never reluctant to give voice to his opinion. Declaring before an 1840 parliamentary select committee that "it would be as well to abolish the office of coroner, as to allow magistrates to exercise any such power" over the inquest's terms of inquiry, Wakley explained that magistrates sought this power because they "are the controlling authorities in the gaols and in lunatic asylums; they are sometimes concerned in cases where life is lost in conflicts between the people and the civil powers; the magistrates are the persons to whom the poor apply in cases of urgent necessity, when the requisite aid is refused to them by parochial officers . . . If coroners be subject to the control of persons who are thus engaged, seeing the tyranny which might be exercised over them with reference to their accounts," he warned, "they might shrink from the performance of their duty at a time when their most powerful energies should be called into action in the public service." *BPP,* 1840, vol. 14, "Report from the Select Committee appointed to inquire into any measures which have been adopted for carrying into effect in the county of Middlesex, the provisions of the Act 1 Vic. c. 68, and also into any proceedings of the Justices of the Peace in relation to the Office of Coroner in the said County," 456.

14. Speaking in third person of his caution with respect to ordering postmortems, Wakley as *Lancet* editor opined: "A display of indiscretion in the official conduct of a medical Coroner who had been elected for the metropolitan county, would have quickly settled the question in the public mind as to the eligibility of the members of the two professions,—the legal and the medical,—for that appointment." *Lancet* 1 (18 Dec. 1841): 411.

15. Ibid.

16. *Lancet* 1 (3 Dec. 1842): 364.

17. *Lancet* 1 (11 Dec. 1841): 377–81, 380.

18. *Lancet* 1 (18 Dec. 1841): 411.

19. Wakley had equivocated somewhat during the debates leading up to the Anatomy Act, opposing its social iniquities but at the same time insisting on the need for professional access to bodies. He agreed that the exclusive focus on pauper bodies was "founded on the cruelty of making an arbitrary disposition of the bodies of the poor, after their lives shall have been worn out in the service of their task masters," and urged instead that "*all*

unclaimed bodies should be appropriated, without reference to the rank or wealth of the deceased." *Lancet* 1 (14 Mar. 1829): 756. Still, Wakley acknowledged that even this would not shift the disproportionate burden from the poor. Accordingly, Wakley sought means (legislative and otherwise) to stimulate a culture of bequests. In his temperate response to the debate, Wakley parted company with many of his Radical supporters, notably Hunt and Cobbett, who were virulently opposed to the bill. Richardson, *Death,* chs. 6–8, esp. 151–57.

20. After pointing out the need for regular inquest postmortems, Wakley warned of the inevitable backlash that would follow an insensitive imposition of such a regime, arguing that, in a cultural climate in which postmortems implied something untoward about a death, such a policy would not only alienate the public but also, by creating "the very *suspicion* with respect to the nature of the death which it is the express object of the inquest to prevent, or to trace to its last source," ultimately subvert the very purpose of inquiry. *Lancet* 1 (18 Dec. 1841): 411. Emphasis original.

21. *Morning Chronicle,* 14 October 1839, 3.

22. *Morning Herald,* 4 July 1839, 7.

23. *Lancet* 1 (12 Nov. 1842): 261. Emphasis original.

24. Wakley advised his jury: "As matters now stand *post-mortem* examinations tended to cast a stigma on innocent families, for, supposing five or six cases of sudden death should occur in any one neighbourhood, and that the jury should require but one *post-mortem* examination, their doing so would show their suspicion of the honesty of one party, whilst it freed from all guilty imputation all the other parties, who may not be a whit more innocent than the other." *Morning Herald,* 4 July 1839, 7.

25. The Greenwalds concluded from their study of Westminster inquests that the rate of postmortems rose from 17% in 1835–38 to 49.7% in 1865–66. Forbes reports, on the basis of his study of select series of London inquest records, that, in Southwark and the City of London in 1820–29, postmortems were held in 23% of inquest cases and that, in the London franchise district of the Duchy of Lancaster in 1831–83, postmortems were done in 13% of cases involving illness, increasing "erratically" over the course of those five decades. Figures published sporadically by the London County Council (LCC) after it assumed control over most of the London districts in 1888 indicate a significant—and steadily rising—rate of metropolitan postmortems at the end of the century. According to the LCC's findings, postmortems were conducted in 48% of inquest cases in 1894, 56.5% in 1907, 62% in 1918, and 79% in 1930. Maria White Greenwald and Gary I. Greenwald, "Medicolegal Progress in Inquests of Felonious Deaths:

Westminster, 1761–1866," *Journal of Legal Medicine* 2, no. 2 (1981): 193–264, 208; Thomas R. Forbes, "Coroner's Inquisitions from London Parishes of the Duchy of Lancaster: The Strand, Clapham, Enfield, and Edmonton, 1831–1883," *Journal of the History of Medicine* 43 (1988); Forbes, "Crowner's Quest," *Transactions of the American Philosophical Society* 68 (1978): 5–52, 44.

26. William Farr, "Letter to the Registrar General On the Causes of Death in England and Wales," *BPP*, 1857–58, 23:240. Copies of Farr's statement were sent to all coroners and to interested medical practitioners. This was a culminating statement of two decades during which Farr had used his annual letters to the registrar general to draw attention to the subject of inquests. Examples of his earlier observations can be found in his first letter, when Farr complained that too many inquest verdicts were "unintelligible to the registrar," and in his third letter, in which he observed that "the principle utility of the inquest is the security which it affords the public mind," but then warned that inquests in which "the 'cause of death' is not investigated, can neither inspire criminals with dread nor the public with confidence." *BPP*, 1839, 16:77; *BPP*, 1841, sess. 2, 6:56.

27. *BPP*, 1857–58, 23:242.

28. Ibid., 249. Note the similarities between this formulation and that analyzed in ch. 1's discussion of the Benthamite model of publicity (section II).

29. M. J. Cullen's *The Statistical Movement in Early Victorian Britain: The Foundations of Empirical Social Research* (New York, 1975) and John M. Eyler's *Victorian Social Medicine: The Ideas and Methods of William Farr* (Baltimore, 1979) provide informative accounts of the origins and functions of the General Register Office (GRO). For recent discussions, see Simon Szreter, ed., *The General Register Office of England and Wales and the Public Health Movement, 1837–1914* (special ed. of the *Social History of Medicine* 4, no. 3 [1991]), esp. Lawrence Goldman, "Statistics and the Science of Society in Early Victorian Britain: An Intellectual Context for the General Register Office" (415–34), and Szreter, "The GRO and the Public Health Movement in Britain, 1837–1914" (435–63).

30. *BPP*, 1839, 16:73. Farr went on to observe that, like other sciences, medicine was beginning to abandon vague conjecture, to substitute numerical expressions for uncertain assertions: "The prevalence of disease, for instance, is expressed by the deaths in a given time out of a given number living with as much accuracy as the temperature is indicated by a thermometer" (64).

31. M. J. Cullen argues that the inclusion of cause in the information

required of a death was the work of Edwin Chadwick, who saw in the provision a way of placating the medical profession angered by the marginal status accorded medical officers in the new Poor Law services. John Eyler, while finding this argument plausible, also points out that Chadwick had quickly grasped the uses of vital statistics as a means of propagating his gospel of sanitary reform. Cullen, *The Statistical Movement,* 53–54, and "The Making of the Civil Registration Act of 1836," *Journal of Ecclesiastical History* 25 (1974): 55–58; Eyler, *Victorian Social Medicine,* 45. Lawrence Goldman has recently suggested that the Registration Act was originally conceived by its Whig architects as a payback to Dissenters for their support in the Reform debates (allowing as it did for non-Anglican registration of baptisms, marriages, and burials) and not as a cornerstone of a well-planned policy of statistical management of population. Goldman, "Statistics and the Science of Society," 418.

32. Before the 1874 act (37 & 38 Vict., c. 88), deaths certified by *any* qualified medical practitioner—whether "registered" (i.e., licensed by recognized medical schools and licensing bodies) or not—could be entered by the registrar as "certified." After 1874 a distinction was enforced between certificates from registered medical practitioners ("certified deaths") and those from unregistered practitioners (thereafter "uncertified"). For a more detailed explanation of these mechanisms, see Anne Hardy, "'Death is the Cure of All Diseases': Using the General Register Office Cause of Death Statistics for 1837–1920," *Social History of Medicine* 7 (1994): 472–92, 474–76.

33. Act for the Registration of Births, Deaths, and Marriages (6 & 7 Will. 4, c. 86), s. 25.

34. Ibid.

35. This system of referral seems to have been in operation early in the workings of the Registration Act, though it was not formalized until the Registrar General's Rules of 1885. Evidence that such communication between the registrar of a district and the local coroner took place earlier can be found in the 1859 and 1860 parliamentary inquiries into the coroner's office, during which one of the central questions was whether to allow such a practice to continue or whether instead to restrict the notification of coroners to the local police.

36. Dr. George Buchanan (medical officer of health for the London parish of St. Giles), cited in Henry W. Rumsey, *Essays and Papers on Some Fallacies of Statistics concerning Life and Death, Health and Disease* (London, 1875), 171.

37. *Lancet* 2 (8 June 1839): 403–6.

38. William Farr, "On Mortality Registration," printed manuscript, c. 1870, in the Farr Collection, London School of Economics [PR], iii.

39. For more extensive analyses of the theoretical basis of nineteenth-century statistics, see Theodore M. Porter, *The Rise of Statistical Thinking, 1820–1900* (Princeton, 1986), esp. chs. 1–2, and Ian Hacking, *The Taming of Chance* (Cambridge, 1990).

40. See Mary Poovey, "Figures of Arithmetic, Figures of Speech: The Discourse of Statistics in the 1830's," *Critical Inquiry* 19 (winter 1993): 256–76, esp. 260–61. See also Porter, *The Rise of Statistical Thinking,* ch. 6, and Hacking, *The Taming of Chance,* ch. 10, for resistance to the application of statistical analysis to medical therapeutics on the grounds of over-abstraction. For a contemporary discussion of this theme, see "Figures of Arithmetic versus Figures of Speech," *Chambers' Edinburgh Journal,* no. 379 (4 May 1839): 113–14.

41. A. B. Granville, *Sudden Death* (London, 1854), 48–49.

42. "Unlike the public health movement of the 1840s," Roy MacLeod observes, "the major organizational impulse for the State Medicine movement of the 1860s and 1870s was generated by the medical profession itself." Roy M. MacLeod, "The Anatomy of State Medicine: Concept and Application," in F. N. L. Poynter, *Medicine and Science in the 1860's* (London, 1968), 199–227, 213. As Christopher Hamlin argues, however, this was a medicine stripped of its former social and political orientations. In his schema, the broadly humanitarian paradigm of a traditional, holistic, and person-centered approach to medicine—and the political implications entailed by it—was effectively marginalized by a model of public health dominated by the highly abstracted and analytical calculus of political economy. When "medicine" reemerged as a significant element in the disciplinary and epistemological profile of public health in the second half of the nineteenth century, Hamlin states, it did so as "an emasculated form of the stillborn political medicine of a half century earlier." Hamlin, *Public Health and Social Justice,* 47 and esp. chs. 1–3.

43. In Roy MacLeod's words, Rumsey advocated "the division of the country into public health districts, coterminous with registration districts, and the appointment of public medical officers, whose duties would include forensic medicine and public hygiene, who would be debarred from clinical practice and who could communicate with a central department of health through a system of inspectors." MacLeod, "The Anatomy of State Medicine," 205–6. This is no place to provide a detailed account of the state medicine debate at midcentury. In highlighting Rumsey's work I merely suggest the broader context for the call for a system of public certifiers. For im-

portant work on the general field of state medicine in Britain, see, in addition to MacLeod (1968), Smith (1979), Eyler (1979), and Hamlin (1997), S. E. Finer, *The Life and Times of Edwin Chadwick* (London, 1952); C. Fraser Brockington, *Public Health in the Nineteenth Century* (Edinburgh, 1965); and Anthony S. Wohl, *Endangered Lives: Public Health in Victorian Britain* (Cambridge, Mass., 1983).

44. Rumsey, *Essays and Papers,* 162.

45. The Coroners' Society was founded in 1846 with the express purpose of protecting its membership's prerogatives. The society's objection to the proposal of a public certifier appeared in the president's annual letter for 1874 (*Annual Report of the Coroners' Society, 1874*). The Manchester coroner, Edward Herford, was the most outspoken opponent from within the ranks of coroners. In "The Office of the Coroner," he counted among the enemies of popular liberties "the heads of Government Bureaus, such as the Registrar-General, who subordinates the protection of human life to the mere collection of statistics, and to whom it is of no consequence whether they are *correct* or not, so long as they are *precisely expressed* in Latin or English terms, and reducible into 'quarterly tables of the causes of death,'" adding that "upon this very narrow—*purely statistical*—basis the Registrar-General and Dr. Farr have constructed a system for practically superseding the Coroner's duty." Edward Herford, "The Office of the Coroner: A Paper Read before the Manchester Statistical Society" (Manchester, 1877), 11, 34. Emphasis original.

46. Rumsey, *Essays and Papers,* 171–72.

47. In 1879, the proposed system of public certifiers received the endorsement of a parliamentary select committee that convened briefly to consider the differences between the English and Scottish procedures for death inquiry. However, the committee's report was not acted upon legislatively and formed no part of the omnibus 1887 Coroners Act, which codified much of the common law precedent concerning inquest law.

48. For a historical overview of these anxieties and his conclusion that concern regarding premature burial persisted in Britain well into the twentieth century, see Martin S. Pernick, "Back from the Grave: Recurring Controversies over Defining and Diagnosing Death in History," in Richard M. Zaner, ed., *Death: Beyond Whole-Brain Criteria* (Dordrecht, 1988), 17–74.

49. *BPP,* 1893–94, vol. 11, "First and Second Report from the Select Committee on Death Certification; together with the Proceedings of the Committee, Minutes of Evidence, Appendix and Index," 473.

50. Ibid.

51. Ibid., 473–74. The committee's report registered both concern and

familiarity with the account of the inquest that Ogle presented them: "This is an old standing cause of complaint," it observed. "It appears that many coroners conduct their inquiries on the supposition that they are only concerned to ascertain whether death has been caused by violence; and, where death is due to natural causes, that it forms no part of their duty to determine what is the disease to which the death is to be attributed." Ibid., 231.

52. Ibid., 337, comments of Sir Walter Foster. Foster was one of the few medically trained members of Parliament and later became president of the Local Government Board.

53. Ibid., "Second Report of the Select Committee on Death Certification," 204.

54. "Death Certification: Defects of the Present System and Suggestions for Their Remedy: Pt. 2," *British Medical Journal* 2 (1 Dec. 1900): 1581.

55. John Graunt, *Natural and Political Observations Mentioned in a following Index, made upon the Bills of Mortality* (1662; reprint, New York, 1975), 37.

56. Rumsey, *Essays and Papers,* 118.

57. For example, J. F. J. Sykes, the medical officer of health for St. Pancras (London) was asked about the tendency to falsify certificates in delirium tremens, syphilis, and suicide cases. He observed that the "medical attendant has to consult the feelings of the family as well as the good of the State, and I think that works a good deal of harm because euphemisms are used which lead to a wrong classification of the causes of deaths. Deaths from alcohol and deaths from syphilis and things of that kind are rarely classified in their naked truth; and the consequence of that is that they get wrongly classified." *BPP,* 1893–94, 11:275.

58. For a background to the concerns that animated the Committee on Physical Deterioration's hearings, see Geoffrey R. Searle, *The Quest for National Efficiency: A Study of British Politics and Political Thought, 1899–1914* (Oxford, 1971), and Daniel J. Kevles, *In the Name of Eugenics: Genetics and the Uses of Human Heredity* (Berkeley, 1986).

59. The Medico-Political Committee of the British Medical Association (BMA) had in November 1903 called upon the BMA to adopt a resolution endorsing confidential certification. Wellcome Institute Contemporary Medical Archive Centre (hereafter CMAC): SA/BMA/C.485, Medico-Political Emergency Sub-Committee, 4 November 1903. For Horsley's BMA activities, see Stephen Paget, *Sir Victor Horsley: A Study of His Life and Work* (London, 1919).

60. *BPP,* 1904, vol. 32, "Report of the Interdepartmental Committee on Physical Deterioration," 536.

61. The list grew in subsequent years, with the most notable addition being the Public Control Committee (PCC) of the London County Council. In its "Recommendations on Coroner's Law Amendment," the PCC adopted verbatim the Physical Deterioration Committee's determination that certificates should be confidential and their contents never divulged. Organizations like the Medico-Legal Society also heard endorsements of this position. In his 1910 lecture on "The Coroner and His Medical Neighbours," London Coroner W. Wynne Westcott lamented the practice of misrepresenting unsavory causes of death as tending to "render the register an unreliable guide to the state of the nation's health. The remedy lies with Parliament, which should demand an accurate certificate of the cause of death, under penalties for false certification; and should grant secrecy to all certificates of death, and order them to be sent direct to the registrar." *Transactions of the Medico-Legal Society* 8 (1910): 15–26, 17. In 1914 the Executive Committee of the General Council of Medical Education and Registration resolved "that in the interests of correct Vital Statistics and for the relief of medical practitioners in the due performance of their functions, it is desirable that Certificates of the Cause of Death should be treated as confidential communications to the Registration Authorities, and that their contents should no longer be open to public inspection." Public Record Office (PRO): MH55/565, "Memorandum on the Registration Acts," March 1915.

62. *BPP*, 1909, vol. 15, "First Report of the Departmental Committee Appointed to Inquire into the Law Relating to Coroners and Coroners' Inquests: Pt. 2, Evidence and Appendices," 435.

63. Alternatively, Latin could be marshaled to shield the practitioner from the resentment of families. William Berry, the medical officer of health for Wigan, described this use of esoteric discourse in an address to his local medical society: "Latin, I think, enables us to certify a cause of death precisely and briefly, and, again, it is not always wise to let the relatives know the cause of death in plain language; they are better pleased if the name of the disease has a jaw-breaking pronunciation and is a term not easily 'understood by the people.'" *Lancet* 2 (24 Sept 1904): 894.

64. Oddie to Horsley, 24 January 1916, Horsley Papers—Vital Statistics (A45), University College London Archives.

65. In 1920 the newly established Ministry of Health, the Home Office, and the Registrar General's Office held a series of meetings to discuss the merits of secret certification. The doctors from the Ministry of Health voiced broad support for the treatment of death certificates as state documents, reiterating the medical practitioners' sense of being caught between public

and private demands. As a remedy, they proposed a form of certificate that would "be identifiable, but not identified, with the names of the deceased, i.e., should be distinguished only by numbers or symbols." This was resisted by the representatives of the two other offices of state. The Registrar General's Office objected on the grounds of administrative complexity, and the Home Office feared that "any system of supplementary certificates would tend to alter the character of the certificates supplemented," leading to a situation in which "doctors might, in every class of case, tend to save all but bald statements for insertion in supplementary certificates." This would create a system that operated on an avowedly fictional and euphemistic principle, which the Home Office believed both endangered the value of public documents and ignored the "natural desire of relatives to have some information." PRO: HO45/12192/37c, "Memorandum of proceedings at the third conference," 4 November 1920, pp. 2 and 3.

66. Cook to Cox, 23 June 1922, Wellcome Institute CMAC: SA/BMA/C.499.

67. S. P. Vivian to Cox, 11 July 1922, Wellcome Institute CMAC: SA/BMA/C.499.

68. These forms were issued to local registrars with strict guidelines as to their distribution: they were to be supplied only to members of the Royal College of Physicians (RCP), Royal College of Surgeons (RCS), or Society of Apothecaries; medical graduates of a university; or apothecaries legally qualified by having been in practice before 1815. In his directive, the registrar general asked registrars to draw up lists of such practitioners, specifying that "the Names of Medical Men who have retired from practice, of Medical Pupils, Students, or Assistants, and of Chemists and Druggists, not to be inserted. No Names of Quacks, or irregular Practitioners, to be enumerated in this list." On the form itself, a side note read: "If this Form should, by accident fall into the hands of any unqualified practitioner, he is recommended not to fill it up." Circular from the General Register Office, 12 July 1845, reprinted in *BPP*, 1846, 19:265–66.

69. "The Vitiation of Death Statistics," *Lancet* 1 (26 Feb. 1876): 321.

70. Ibid.

71. "The Registrar-General and Unqualified Practitioners," *Lancet* 1 (12 Feb. 1876): 253.

72. *Annual Report of the Coroners' Society,* 1897–98.

73. *Annual Report of the Coroners' Society,* 1905–6, 213. Other organizations joined in. The Medical Defense Union, founded in 1886 as the profession's legal watchdog, spent the first few decades of its existence prosecuting quacks who misrepresented their information to registrars of death

as "certified" information and punishing orthodox practitioners who assisted in the deception by "covering." The BMA's Medico-Political Committee issued a statement in 1903 that "the present death certification and registration renders unreliable the statistics of death from disease, tends to encourage and assist quack practitioners in their impositions upon the public, and facilitates the concealment of crime." Wellcome Institute CMAC: SA/BMA/C.485, Meeting of the Emergency Sub-committee, 4 January 1903. For a history of the Medical Defense Union's response to such breeches in professional standards of conduct, see Robert Forbes, *Sixty Years of Medical Defense* (London, 1947). For a detailed analysis of the developing system of professional conduct, which includes a revealing appendix listing the cases brought to the attention of the General Medical Council, see Russell G. Smith, *Medical Discipline: The Professional Conduct Jurisdiction of the General Medical Council* (Oxford, 1994).

74. *British Medical Journal* 1 (15 Feb. 1902): 431. The 25 February 1905 issue reported on another instance of this technique. The West Lancashire coroner, S. Brighouse, at an inquest on a herbalist's patient, warned the herbalist "that if he attended other people, and attempted to diagnose their complaints, and gave certificates in the event of their death, an inquest would be held in every case." *British Medical Journal* 1 (25 Feb. 1905): 432.

75. PRO: MH58/106, "Amendment of the Registrar General's Regulations for Registrars of Births and Deaths, Regulation No. 21 . . . and Regulation No. 25" (1914), 5.

76. For a discussion of the tension between orthodox medicine and the "conscientious objection" of groups like the Christian Scientists, see Roy MacLeod, "Medico-Legal Issues in Victorian Medical Care," *Medical History* 10, no. 1 (Jan. 1966): 44–49.

77. PRO: RG48/430, minutes of meeting at the GRO, 7 May 1925.

78. Charles W. J. Tennant to the lord chancellor, 24 November 1926, PRO: HO45/24752/427806/23.

79. PRO: HO45/11214/403923/13, 30 July 1919. The Home Office and the Registrar General's Office were not unmoved by such protests. In an internal memo on the Tennant and Culpan cases, Home Office clerk Herbert B. Simpson urged support for unorthodox practitioners:

The law does not prohibit anyone from taking to herbalism, Christian Science or any other mode of treating a disease which is not recognised by the medical profession . . . There seems to me to be no doubt at all that the implicit faith in medical science which was so prevalent a generation ago has been greatly weakened in recent years, partly by the admissions of the medical practitioners themselves. I think the Association might be assured that the Secretary of State does not approve the very hostile attitude which some of the Coroners are alleged to have adopted towards them.

HO45/24752/427806/13, 30 June 1925.

Chapter 3: From the Alehouse to the Courthouse

1. S. Squire Sprigge, *The Life and Times of Thomas Wakley* (London, 1889), 354.

2. The London County Council (LCC) took over administrative control of London inquests after the 1888 Local Government Act (51 & 52 Vict., c. 41). Its campaign against pub inquests was statutorily supported by the 1891 Public Health Act for London (54 & 55 Vict., c. 76), which required every metropolitan sanitary authority to provide a mortuary and the LCC to provide and maintain proper accommodation for holding inquests. Despite resolute efforts at compliance, 583 of the 7,300 metropolitan inquests in 1896 were still held in pubs, albeit largely confined to a single South London subdivision. With the completion of a new coroner's court for this district in 1897, the figure plummeted to 28, and by 1899 the LCC congratulated itself and the people of London for having closed the book on the metropolitan pub inquest.

3. Olive Anderson, *Suicide in Victorian and Edwardian England* (Oxford, 1987), 33n. 69. Forbes's study of Middlesex inquest records for the first half of the nineteenth century found that over 80% of inquests were held in pubs. T. R. Forbes, "Coroners' Inquests in the County of Middlesex, England, 1819–42," *Journal of Medical History* 32 (Oct. 1977): 377. My sampling of metropolitan inquest records suggests that this figure holds for the second half of the century as well. Pubs served as venues for 88% of inquests in the Western district of Middlesex in January to June 1856; the figure in the same district for 1870 was 84%. London Metropolitan Archives (LMA): COR/A/1, 1856; COR/A/4, 1870. In the final quarter of the century, however, the site of inquests was clearly in transition, a fact that is quite literally inscribed on the printed inquisition forms used by coroners. Middlesex coroners were still using forms that began with the words: "An Inquisition taken for our Sovereign Lady the Queen, at the House known by the Name of ____," with a space left for the coroner to insert "The Royal Oak" or some such venerable local establishment. By the late 1880s, however, coroners had begun to strike out "at the House" and insert "at the Coroner's Court" or "at the Town Hall" in its place.

4. Public Record Office (PRO): HO45/17009/533786/24A&24B. Replies to the letter from the Coroners Committee to Borough Chief Constables, 28 June 1835.

5. Brian Harrison, *Drink and the Victorians: The Temperance Question in England, 1815–1872* (London, 1971), 55.

6. To be sure, radicals in the first half of the nineteenth century were split as to the propriety of tavern-centered politics, with many important figures—the rationalist Richard Carlile and the dogged moral improver Francis Place most emphatically—condemning the excesses of popular political culture, of which the tavern meeting was a prime example. Wakley was not one of these critics. In the nearly forty-year span of his wide-ranging and often aggressively opinionated editorship of the *Lancet,* he voiced no criticism of the pub inquest. Indeed, where Wakley's editorials did refer to the physical context for inquests, they endorsed its popular features: The "roving character of the office" ensured that "the operation of inquests is not confined to palaces and mansions" and that "no difficulties or offensiveness of place or locality" would bar the proper execution of its duties (*Lancet* 1 [12 Nov. 1842]: 260). For important discussions of the tension between the "respectable" and "carnivalesque" faces of radicalism, see Iain McCalman, *Radical Underworld: Prophets, Revolutionaries, and Pornographers in London, 1795–1840* (Oxford, 1993), esp. chs. 6 and 9, and James Epstein, *Radical Expression: Political Language, Ritual, and Symbol in England, 1790–1850* (New York, 1994), ch. 4.

7. I use the term *profane* in the senses both of "not participating in or admitted to some esoteric knowledge" and of things regarded as desecrating and "ritually unclean or polluted" (*Oxford English Dictionary*). Much of the medical criticism of the inquest resonates with this double meaning. Mary Douglas's *Purity and Danger: An Analysis of Concepts of Pollution and Taboo* (New York, 1966) provides the classic framework for analyzing these physical and conceptual boundaries. For a related discussion in the context of nineteenth-century English public health, see Peter Stallybrass and Allon White, *The Politics and Poetics of Transgression* (Ithaca, N.Y., 1986), ch. 3, "The City: The Sewer, the Gaze, and the Contaminating Touch."

8. Christopher Hamlin's study of the Victorian public health movement provides an important and vigorously argued version of this analysis, one that is particularly relevant here in that it places medical radicals like Wakley and their assorted allies squarely on the "losing side" of the great struggle to define the approach to public health in the pre-Chadwick era. Ham-

lin, *Public Health and Social Justice in the Age of Chadwick* (Cambridge, 1998), esp. chs. 1–3.

9. Charles Dickens, *Bleak House,* ch. 11, "The Sol's Arms." Dickens was himself a juror at an inquest presided over by Wakley and wrote favorably of his experience with the famed coroner in *The Uncommercial Traveller,* ch. 18 of *Works of Charles Dickens* (New York, 1926), 13:195–99.

10. "A Coroner's Inquest," *Household Words* 1 (1850): 109–13, 109, 110.

11. Ibid., 110. This article drew a sharp response from Joshua Toulmin Smith, one of the outspoken champions of the inquest as a bastion of local liberty. In his 1857 edition of *The Parish,* Smith denounced "superficial persons"—and pointed the finger directly at the "very discreditable tone of the observations in a paper in Dickens's *Household Words*"—who, seeing only "the external tinsel of the administration of justice, and not its deep reality, have often sought to ridicule the Coroner's Inquest" for its informal setting. This was, however, one of the most admirable parts of the institution, Smith insisted. "It has no cumbersome and costly machinery, which can only be worked by functionaries at some mysterious central abode. It brings the eye of the Law directly home to every spot; so that no fact shall escape the searchingness of direct inquiry, and that every man shall know and feel the immediate presence of the course of Justice." Joshua Toulmin Smith, *The Parish: Its Powers and Obligations at Law,* 2d ed. (London, 1857), 380.

12. *Hansard Parliamentary Debates,* 3d ser., 230 (June–July 1876): 1305.

13. John Belchem and James Epstein, "The Gentlemen Leader Revisited," *Social History* 22, no. 2 (May 1997): 174–93, 186. Belchem and Epstein's article provides a critical assessment of a vibrant and growing literature on this subject. For the present discussion, most salient is James Vernon's challenging reconsideration of the mid-Victorian "politics of culture." Vernon, *Politics and the People: A Study in English Political Culture, c. 1815–1867* (Cambridge, 1995), esp. ch. 6, which contains a discussion of the assault on the public house as a venue for radical politics. Also relevant, at a self-consciously more abstracted level, is Mary Poovey, *Making a Social Body: British Cultural Formation, 1830–1864* (Chicago, 1995), ch. 2.

14. "An Inquest-Room for the City," *British Medical Journal* 1 (14 Apr. 1877): 460.

15. John Payne, coroner for the City of London and Southwark, to the commissioners of sewers, 12 March 1877, Guildhall Library Misc. MSS 149.8.

16. PRO: HO45/9680/A47890E/1, "Petition from the Mayor, Aldermen, and Burgesses of the County Borough of Sheffield," 23 September 1889.

17. *The Times*, 13 October 1895, 7.

18. Albert B. Deane, ed., *The Licensed Victuallers' Official Annual, Legal Textbook, Diary and Almanack for the Year 1906* (London, 1907), 211.

19. *Annual Report of the Coroners' Society*, 1895.

20. 29 & 30 Vict., c. 90. The enabling clause meant that the government permitted, but did not require, the levying of public funds by local authorities for the purpose.

21. "A Mortuary House for Westminster," *Sanitary Record* 5 (11 Nov. 1876): 313. Over the next few years, the case for mortuary construction on a national scale became a familiar sanitarian topic. See, e.g., Henry Burdett's plea for reform as a means of ending "the present indecency of inquests." Henry Burdett, "On the Importance of Mortuaries for Towns and Villages," reprinted in *Sanitary Record*, n.s., 2 (15 Oct. 1880): 130–37, 134.

22. "Extracts from the Minutes of the Commissioners of Sewers relative to the Provision of a Coroner's Court Room," Report of the Medical Officer of Health, 9 January 1877, Guildhall Library, Misc. MSS 149.8.

23. John Payne, coroner for the City of London and Southwark, to the commissioners of sewers, 12 March 1877, Guildhall Library, Misc. MSS 149.8.

24. "Inquest Courts," *Lancet* 1 (19 Jan. 1878): 100.

25. London Metropolitan Archives (LMA): LCC/PC/COR/1/111, "London County Council. Mortuaries and Coroners' Courts: Model Plans and Suggestions," 1892, 1.

26. Extract from the *Paddington Mercury*, in LMA: LCC/PC/COR/3/26, PCC Report on the Paddington Coroner's Court, 1902.

27. LMA: LCC/PC/COR/1/111, "Model Plans," 1892, 1.

28. LMA: LCC/PC/COR/1/111, "Report of the Chief Officer of the Public Control Department as to the provision made for Coroners' Courts, Mortuaries, and Post-Mortem Chambers in each of the new London Municipal Boroughs," January 1900, 10.

29. Letter from estate manager of Lord Llayetto, 20 March 1895, LMA: LCC/PC/COR/3/6, Public Control Committee files on London Coroners' Courts and Mortuaries, Camberwell District.

30. Hammersmith District, solicitor's letter to the Hammersmith Vestry, 13 June 1894, LMA: LCC/PC/COR/3/14.

31. "The Condition of Bermondsey's Public Mortuary," *Southern Recorder*, 12 January 1889. This concern with seclusion and privacy shared

the same conceptual space as the longstanding discussion about the proper context of anatomical dissection. The overlap in imagery and sensibility is apparent in the 1877 case of "a private dissecting-room" reported in the *Lancet*. An outraged member of the public had brought suit against Thomas Cooke, a medical man, charging Cooke with endangering public health and morals by operating a dissection chamber in full public view. The *Lancet* editors viewed the case with regret:

> It is sufficiently obvious that popular objections against the idea of a dissecting-room and its contents being next door neighbours to one would be very strong, and we can only regard Mr. Cooke's want of consideration for public opinion as rather unfortunate. It would be absurd to consider a dissecting-room detrimental to health, but it has objectionable features, and should be detached from other premises, have a private entrance, and not be a house taken promiscuously in a public street. In the latter case, in addition to many other objections, it is almost impossible to ventilate it sufficiently, and without offence to those who live adjacent, and at the same time to screen the operations that must be carried out therein from the prying gaze of curious neighbours.

"A Private Dissecting-Room," *Lancet* 1 (20 Jan. 1877): 101. Cooke defended himself in the next week's issue, claiming to have performed an experiment proving that neighborhood offense was based on prejudice alone. He had friends bring an empty coffin to his door, at which time his neighbors, "on seeing an empty coffin brought in, and believing that there was a body in it, they again said 'they perceived the smell as before.'" *Lancet* 1 (27 Jan. 1877): 152.

32. Viewing a body, of course, had connotations in popular mortuary practice as well. See Ruth Richardson, *Death, Dissection, and the Destitute* (London, 1988), 22–29.

33. Sir John Jervis, *A Practical Treatise on the Office and Duties of Coroners* (London, 1829), 28. The 1819 Peterloo inquest was quashed on grounds that the coroner did not view the body, and this politically controversial ruling became the unquestioned judicial ground for its inviolability.

34. Ibid., 27–28.

35. Lord Chancellor's Office to Home Office, 24 November 1889, PRO: HO45/9680/A47890E/2.

36. *R. v. Haslewood* (1926). The provisions of the 1926 act were that the jury's view was no longer a requirement but could be conducted if a majority of the jurors wished to see the body. The coroner's view remained

mandatory until World War II and became discretionary afterward. According to coroners' responses to a Home Office survey of inquest proceedings after the 1926 act, no jury ever exercised their discretionary privilege. The 1926 act is discussed in some detail in the epilogue of this book.

37. Charles Rothera, coroner for Nottingham, in *Annual Report of the Coroners' Society,* 1894, 9.

38. *Hansard,* 4th ser., 187 (Mar.–May 1908), 862.

39. Norbert Elias, *The Civilizing Process,* trans. Edmund Jephcott (New York, 1978), vol. 1. This double-edged reading informs the other obvious framework within which to situate calls to shut down public spectacles like public execution—Michel Foucault's analysis of the birth of a modern disciplinary form of power—though, of course, Foucault's immediate theoretical inspiration is more Nietzsche than Freud. Foucault, *Discipline and Punish: The Birth of the Prison,* trans. Alan Sheridan (New York, 1977); Friedrich Nietzsche, *Genealogy of Morals,* trans. Walter Kaufmann and R. J. Hollingdale (New York, 1989), Second Essay.

40. At a literal level, by analogizing the view of the body to public execution as grounds for its elimination, Higham was simply pressing the historical process of displacement one step further: Just as the transparency attributed to the coroner's jury allowed an absented public to "know" that an execution had been properly carried out, so, too, might the inquest jury, by recourse to medical expertise as a representative embodiment of reliable knowledge, be provided with any necessary information to be had of a physical body under inquiry. As will be suggested, however, this further "refinement" entailed its own complex (and uneven) logic of compensation.

41. "Petition from the Mayor, Aldermen and Burgesses of the County Borough of Sheffield," 23 September 1889, PRO: HO45/9680/A47890E/1.

42. This was the case despite the fact that, as the *Lancet* pointed out in 1889, "the question as to how far a corpse can be considered infectious is one concerning which very considerable difference of opinion exists." "The Infectivity of a Corpse," *Lancet* 1 (15 June 1889): 1206. Tom Laqueur has found that early English sanitarians like Southwood Smith, for whom the unregulated interplay of the living and the dead served as the exemplary image illustrating the need for reform, themselves doubted the pathogenic virulance attributed to the corpse in the discourse of public health. Laqueur, "The Places of the Dead in Modernity," in Dror Wahrman and Colin Jones, eds., *The Age of Cultural Revolutions* (Berkeley, 2000). I thank Prof. Laqueur for permission to cite this article prior to publication.

43. PRO file HO45/10110/B10886 contains several petitions dating from the first decade of the twentieth century.

44. LMA: LCC/PC/COR/3/54, advertisement sent by William C. Thackray, Engineer, of London (SW), undated.

45. Ibid., distributed by Thomas Christy and Co. of London (EC). Those interested in a demonstration of the promised purified view were directed to the City of London Mortuary, which boasted the first De Rechter in England.

46. "The Abolition of 'The View' at Inquests," *British Medical Journal* 2 (1 Oct. 1898): 995–96.

47. William Baker, *A Practical Compendium of the Recent Statutes, Cases, and Decisions Affecting the Office of the Coroner* (London, 1851), 382. Reprinted from Baker's letter to the registrar general, 17 July 1840. The context for these remarks was the acrimonious struggle with county magistrates over inquest payments discussed in ch. 2, sect. I.

48. "The Coroner's Inquest Bill," *Lancet* 2 (18 July 1908): 168. In its assessment of a bill drafted in 1908 abolishing the jury's view, the Home Office agreed that the view, on an evidentiary level, was an anachronism. Herbert Simpson, the Home Office's resident expert on matters relating to inquests, doubted "whether now-a-days the view is ever regarded as an important element in the jury's consideration—they probably are usually ready to take one 'expert' opinion rather than the evidence of their own senses." PRO: HO45/23983/138729/12, undated memo of Herbert B. Simpson, written on the 1908 Home Office docket monitoring parliamentary consideration of a bill eliminating the jury view.

49. *Durham Chronicle,* 16 July 1897.

50. *British Parliamentary Papers* (hereafter *BPP*), 1909, vol. 15, "First Report of the Departmental Committee Appointed to Inquire into the Law Relating to Coroners and Coroners' Inquests: Pt. 2, Evidence and Appendices," evidence of Mr. G. Bateman, Secretary of the Medical Defense Union, 2 April 1909, 573.

51. *Annual Report of the Coroners' Society,* 1897–98, 48.

52. "Public Control Committee Report," 29 January 1895, 2, LMA: LCC/PC/COR/1/1.

53. *BPP,* 1909, testimony of John Graham, coroner for the Chester Ward of County Durham, 15:601.

54. *British Medical Journal* 2 (1 Oct. 1898): 995.

55. E. M. Harwood, "On Coroners" (Bristol, n.d.), 16. (Also excerpted in *Solicitors' Journal* 29 [18 Apr. 1885]: 400.)

56. R. Henslow Wellington, *The King's Coroner, being the Practice and Procedure in His Judicial and Ministerial Capacities* (London, 1906), 2:17. Not everyone agreed that a view in such cases was an unquestioned good.

Though conceding that the view might be useful in cases of emaciation, the president of the Coroners' Society warned his members to be "careful that no sentiment is imparted into the inquiry" (1897). William Brend, in his thoughtful "Amendment of the Law of Coroners," was more skeptical still. Even in cases of marked external violence, the standard surface view was inadequate: "To be of any value the jury should see not only the surface of the body, but also the injuries produced in the internal organs." William A. Brend, "The Necessity for Amendment of the Law Relating to Coroners and Inquests," *Transactions of the Medico-Legal Society* 10 (1912–13): 143–97, 178.

57. *BPP*, 1910, vol. 21, "Second Report of the Coroners' Committee: Pt. 2, Evidence and Appendices," 640.

58. Ibid., 21:8.

59. PRO: HO45/23983/138729/12 (1908).

60. The image problem of coroners was a longstanding concern, with inquests depicted as a joke in literary and legal text alike. (The derisory reference to "Crowner's quest law" in *Hamlet* is only the most famous of the literary slights directed at the inquest.) In the late nineteenth century, the coroner as a risible figure was certainly something of a stock image: Frederick W. Lowndes, MRCS, concluded his paper on inquests delivered at the Liverpool Medical Institute in November 1876 with an apology that his remarks were not as serious as the subject merited: "My apology must be that, by a strange contradiction, it is one of the peculiarities of the coroner's court in this country, that it never can be discussed without exciting ridicule." *London Medical Gazette* 2 (20 Oct. 1877): 432.

61. *Annual Report of the Coroners' Society*, 1905, 183. Comments of E. A. Gibson on the proposed motion for the adoption of coroners' robes.

62. Ibid., 184. Comments of John Troutbeck, coroner for South West London. Troutbeck is discussed at length in ch. 4.

63. Ibid., 185. Comments of R. M. Mercer, Kent County coroner.

64. S. Ingelby Oddie, *Inquest: A Coroner Looks Back* (London, 1941), 14.

65. In a poll of coroners conducted in preparation for the 1909 Coroners Committee, over half thought that the view should be discretionary by the coroner and jury, while only 17% thought that it should be mandatory for both. PRO: HO45/23983/138729/12.

66. *Annual Report of the Coroners' Society*, 1897–98, 45.

67. *BPP*, 1893–94, vol. 11, "Reports from the Select Committee on Death Certification," 331.

68. *BPP*, 1910, 21:658. As late as 1924 the *Manchester Guardian*

reported on a public disturbance resulting from the city coroner's insistence on removing a body to the municipal mortuary.

69. George B. Nadler, Southampton town clerk, to William Coxwell, borough coroner for Southampton, 17 January 1896, *Annual Report of the Coroners' Society,* 1895–96, 22.

70. Ibid., 24.

71. *BMJ* 2 (1 Oct. 1898): 995–96.

Chapter 4: Telling Tales of the Dead

1. Douglas Maclagan, "Forensic Medicine from a Scotch Point of View" (paper presented at the Forty-sixth Annual Meeting of the British Medical Association, Bath, Aug. 1878), reprinted in *British Medical Journal* (hereafter *BMJ*), 2 (17 Aug. 1878): 235–39, 237. Maclagen was professor of medical jurisprudence at the University of Edinburgh, and his deep involvement with the Scottish system of death inquiry made him a strong critic of English practice.

2. "Death Certification and Amendment of the Coroners' Act," *Lancet* 1 (13 Apr. 1907): 1027–28.

3. "Act to provide for the Attendance and Remuneration of Medical Witnesses at Coroners Inquests" (6 & 7 Will. 4, c. 89, s. 7 and s. 2).

4. 6 & 7 Will. 4, c. 89, s. 1. In defending the bill against criticism that it violated the principle that testimony was a civic responsibility devolving upon all members of the community, Wakley and his supporters invoked precisely this distinction between proximity (lay) and skill (expert). An ordinary witness was called, Wakley explained in a *Lancet* editorial, "because he was accidentally cognisant of the facts on which he is to be examined. In the other case the witness attends officially, and the value of his testimony depends on the time and money he may have devoted, in order to acquire a scientific knowledge of his profession." *Lancet* 1 (13 Feb. 1836): 800.

5. Alfred Swaine Taylor, *Elements of Medical Jurisprudence* (London, 1836), 1. There is a large historical literature echoing the laments of Taylor and his contemporaries about the "retarded" development of English medical jurisprudence. See, e.g., J. D. J. Havard, *Detection of Secret Homicide: A Study of the Medico-legal System of Investigation of Sudden and Unexplained Deaths* (London, 1960); Erwin Ackerknecht, "Legal Medicine in Transition," *Ciba Symposia* 11, no. 7 (winter 1950–51): 1290–98; and Thomas R. Forbes, "Early Forensic Medicine in England: The Angus Murder Trial," *Journal of the History of Medicine and Allied Sciences* 36 (1981): 286–309. For a critical overview of this literature, see both Michael

Clark and Catherine Crawford's introduction to their edited collection *Legal Medicine in History* (Cambridge, 1994), 1–17, and Crawford's contribution, "Legalizing Medicine: Early Modern Legal Systems and the Growth of Medico-legal Knowledge" (89–109).

6. Alfred Swaine Taylor, *The Principles and Practice of Medical Jurisprudence* (London, 1865), xx.

7. *London Medical Gazette* 19 (3 Dec. 1836): 344. Emphasis original.

8. This was the case despite ostensible progress in training. As Michael Clark and Norman Ambage have recently explained, "Although attendance for two terms at a course of lectures on 'medical jurisprudence' had been a compulsory part of medical curriculum in England and Scotland since the 1860s, there was no standard course of studies, the quality of instruction often left much to be desired, and scarcely any opportunities for students to gain practical experience in the conduct of medico-legal post-mortems." Michael Clark and Norman Ambage, "Unbuilt Bloomsbury: Medico-legal Institutes and Forensic Science Laboratories in England between the Wars," in Clark and Crawford, *Legal Medicine*, 293–313, 293–94.

9. Early calls for an acceptance of Continental-styled "experts" were made by the University of London's tireless advocate for medical jurisprudence, John Gordon Smith: "There ought, in fact, to be some provision by law, for furnishing adequate aid to courts of justice, to make it the business (and why not the interest?) of unexceptionable *experts*, as the French designate them, to apply themselves to that duty." Smith, *An Analysis of Medical Evidence: Comprising directions for practitioners, in the view of becoming witnesses in courts of Justice* (London, 1825), 173. Emphasis original.

10. "Pathology as a Specialty," *Lancet* 2 (20 Oct. 1877): 579–80.

11. The transformation in pathology was decidedly more equivocal in practice, however. While most of the London medical schools had offered some form of instruction in pathology and morbid anatomy since the late 1850s, these courses were commonly taught as a subsidiary part of other subjects, a situation reflected in the General Medical Council's omission of pathology as one of the ten subjects it proposed for a minimum curriculum in 1867. Teaching arrangements became more secure in the 1870s, but the facilities for specialist research and teaching remained elusive. Even in large teaching hospitals, laboratories were introduced only at the turn of the century. For developments in English pathology during the closing decades of the nineteenth century, see Russell Maulitz, "The Pathological Tradition," in W. F. Bynum and Roy Porter, eds., *Encyclopedia of the History of Medicine* (London, 1993), 169–91; George J. Cunningham and G. Kemp

214 *Notes to Pages 111–113*

McGowan, *History of British Pathology* (Bristol, 1992); J. H. Dibble, "A History of the Pathological Society of Great Britain and Ireland," *Journal of Pathology and Bacteriology* 73 (suppl.): 1957, 1–35; and W. D. Foster, "The Early History of Clinical Pathology in Great Britain," *Medical History* 3 (1959): 173–87.

12. *The Times*, 20 March 1876, 9, and 15 October 1877, 9.

13. Alfred Swaine Taylor, *The Principles and Practice of Medical Jurisprudence*, 2d ed., 2 vols. (London, 1873), 1:14.

14. Farrer Herschell, "Address on Jurisprudence, and Amendment of the Law," *Transactions of the National Association for the Promotion of Social Science*, Liverpool Meeting, 1876 (London, 1877), 22–43, 26–32.

15. "Report of the Select Committee on the Coroners Bill," in *Annual Report of the Coroners' Society*, 1879, 3.

16. *Lloyd's Weekly Newspaper*, 21 October 1877, 1; *Penny Illustrated Paper*, 13 October 1877, 226.

17. PRO: HO144/26/64091/4, petition to Secretary of State R. A. Cross.

18. The Home Office inquiry called on the advice of several leading medicolegal figures, including Taylor. Taylor's opinion underscores the complexity of the broad issues involved in the postmortem question. While advocating a reformed system based on expertise, Taylor was not willing to completely marginalize the general practitioner (one of whom in the Staunton case had been his student some twenty years before). He advised that protests about the inadequacy of the postmortem conducted by the Penge doctors—notably the lack of microscopic examination—were groundless (hardly any inquest examination ever involved the microscope). He also thought that the suspicion of the natural disease as the cause of death would have been best assessed by the general practitioners involved; postmortem appearances might deceive, he observed, and thus "it would be safer to rely upon the opinions of those who saw and described it than upon those who did not see it." PRO: HO144/26/64091/27, Alfred Swaine Taylor, "Draft Report," 5.

19. "Reform of the Coroner's Court," *Lancet* 2 (22 Dec. 1877): 930.

20. See ch. 2, n. 25.

21. The first English edition appeared in 1885, but Virchow was well known in England before then. All references in this work are to the third American edition of T. P. Smith's English translation: Rudolph Virchow, *Post-Mortem Examinations with Especial Reference to Medico-legal Practice*, trans. T. P. Smith (Philadelphia, 1896).

22. Ibid., 12–13. In this passage Virchow was quoting from a medical school lecture he had delivered in 1859. See Lester S. King and Marjorie C.

Pathology 73,
's efforts to

cept.

d-

ren-

after the

he cavity,

ugh.'" Eng-

ics, however.

he pathologist to

amined first, while

and morbid anatomy

oponents of microscopic

y be examined in any order,"

where time is allowed." Beale,

king Post-mortem Examinations

athological Society of London Minute

of the founding meeting of the Society, 8

, "An Address on the Study of Morbid Anatomy,"

iological Society of London 49 (1897–98): il–lxii,

on for pathology, of course, was not without its crit-

o years after Payne's robust defense, his successor to the

ed frustration with the society's "judicious" restrictions,

that "the simple exhibition of specimens is merely making our

eptacles for facts" ("President's Address," Transactions 51 [1900]:

11). The call for gross pathology to yield its place to more subtle,

amic, and therapeutically fruitful fields like chemical and physiological

athology was a common refrain during this period, but this was not a trend
that found much of an echo in the medicolegal literature. For those who
craved clear results, an author of a leading pathological textbook somewhat
ruefully acknowledged, gross pathology prevailed over the more experimental avenues of pathological work: "Morbid anatomy has ever been a

kers in our profession," T. H. Green
of Pathology at the British Medical
is natural that it should be so," he
ble of demonstration, and for the
n are the other methods of inves-
30). Those arguing the case for
examinations would bring to
generally, were understand-
athological knowledge with
s that they held to be the
d judgment.
the Central District of
e Central District of
ortem debate as it
the Medico-Legal
inical knowledge
less knowledge
, were elusive
n a medical
quite com-
tible with
e accord-
entries
ations
dico-

favourite study with the junior wo[...]
observed in his address to the Sectio[...]
Association's 1883 annual meeting. "I[...]
continued, since this work "is more cap[...]
most part beset with fewer difficulties, th[...]
tigation" (*BMJ* 2 [4 Aug. 1883]: 229–32, [...]
expert postmortems and the clarity that suc[...]
the medicolegal field, as Green suggests mor[...]
ably eager to keep intact the identification of [...]
the materially solid, static anatomical specime[...]
pathologist's proper objects of contemplation an[...]

27. Edwin Lankester, *Second Annual Report [...]
Middlesex* (London, 1865), 196.

28. Edwin Lankester, *Sixth Annual Report of [...]
Middlesex* (London, 1869), 16.

29. This problem featured prominently in the post[...]
unfolded in subsequent decades. F. J. Smith argued befor[...]
Society that the uncritical interplay between previous c[...]
and postmortem investigation produced an illusion of sea[...]
concerning the cause of death. Causes of death, Smith argue[...]
in reality and often inaccessible. One of the reasons why thi[...]
ter appreciated, he informed the 1906 meeting, was that "wh[...]
man who has seen the patient during life does the autopsy he is[...]
petent to say that what he found post-mortem was quite compa[...]
the symptoms he has observed during life, and can label the diseas[...]
ingly." Labeling on this basis, Smith maintained, favored fictitious[...]
into the nation's death ledgers. F. J. Smith, "Post-Mortem Examin[...]
Which Do Not Reveal the Cause of Death," *Transactions of the Me[...]
Legal Society* 3 (1905–6): 37–54, 39.

30. *BMJ* 2 (19 July 1902): 232.

31. Taylor, *Principles* (1873), 1:8, 1:31. This advice carried into twent[...]
eth-century texts. W. G. Aitchison Robertson, for example, warned in his[...]
section on "Rules to Be Observed in Giving Evidence" that, as the participants in public inquiries were not medically skilled, "ordinary language
should be employed. 'A tumefaction over the malar region' hardly expresses
to the layman the presence of a swelled cheek, nor does 'a punctured wound
in the left inframammary region' convey the knowledge that there was a
stab in the left breast." W. G. Aitchison Robertson, *Manual of Medical
Jurisprudence, Toxicology, and Public Health* (London, 1908), 15.

32. H. D. Littlejohn, "On the Practice of Medical Jurisprudence," *Edinburgh Medical Journal* 21, no. 1 (1875): 385–93, 392.

33. Stanley B. Atkinson, *Golden Rules of Medical Evidence* (Bristol, n.d.), 32. Atkinson, a London-based barrister and justice of the peace, was a founding member of the Medico-Legal Society and served as its first honorary secretary and co-editor of its *Transactions*.

34. "By the observance of decorum," Smith explained, "I do not mean that we should clothe ourselves in a particular garb, or wear an especial countenance, or affect any occasional fashion of speech or behaviour . . . Let us go forward in our natural character, armed with sobriety and circumspection, and adorned with dignity and modesty." Smith, *Analysis*, 101–2.

35. Robertson, *Medical Jurisprudence*, 15–16; Atkinson, *Golden Rules*, 30, emphasis original.

36. H. H. Littlejohn, "Medico-Legal Post-Mortem Examinations," *Transactions of the Medico-Legal Society* 1 (1902–4): 14–29, 27.

37. Atkinson, *Golden Rules*, 18.

38. Joan Margaret Ross, *Post-mortem Appearances*, 3d ed. (Oxford, 1937), 5–6.

39. G. Sims Woodhead, *Practical Pathology: A Manual for Students and Practitioners*, 4th ed. (London, 1910), 4–5.

40. Ibid., 6.

41. H. D. Littlejohn, "On the Practice of Medical Jurisprudence," *Edinburgh Medical Journal* 21, no. 2 (1876): 1112–24, 1115.

42. Ibid., 1115.

43. W. J. Collins served as chairman of the Public Control Committee (PCC) from 1895 to 1904. He also promoted the cause of medicolegal reform in other capacities, notably as the first president of the London-based Medico-Legal Society and, after 1906, as a member of Parliament for St. Pancras. For the history of the London County Council (LCC) in this period, see John Davis, *Reforming London: The London Government Problem, 1855–1900* (Oxford, 1988), chs. 5–6; Ken Young and Patricia L. Garside, *Metropolitan London* (London, 1982), chs. 3–5; and Susan Pennybacker, "The Millenium by Return of Post: Reconsidering London Progressivism, 1889–1907," in Gareth Stedman Jones and David Feldman, eds., *Metropolis London: Histories and Representations since 1800* (London, 1989), 129–63.

44. London Metropolitan Archives (LMA): LCC/PC/COR/1/1. PCC Report, 29 January 1895, 2.

45. This was, in one sense, a simple organizational task. In 1893 the

PCC had already set up an informal list of pathologists and toxicologists willing to conduct inquest postmortems in "special cases," including some of the best-known figures in the London medicolegal community: Thomas Bond, a surgeon to the metropolitan police, and Drs. Luff and Pepper, both lecturers in forensic medicine at St. Mary's Hospital, the leading medicolegal institution in London at the turn of the century. An attempt to expand and formalize the list a decade later, however, received a lukewarm reception from the hospital establishment. In a joint response to a PCC circular for names of possible participants, the leading London hospitals and the Royal Colleges informed the council that they "fully appreciate[d] the efforts . . . to provide Coroners with competent pathological assistance" and drew up a list in accordance with the request. They insisted, however, that, in the interest of both the public and the profession, pathologists' expertise should be sought "only in those exceptional cases in which expert advice is really necessary" and that the experts should, "save under the most exceptional circumstances, have the co-operation and assistance of the medical practitioner who was in charge of the deceased person during life, or who was the first to be called in after death" (London Hospital Archives, LM/1/4, 253–58). In the contemporary climate of tension between general practitioners and hospitals over proprietorial rights to patients, the hospital administrators were clearly reluctant to pick a fight that held out so little in immediate benefit. Administrators were also keen to ensure that their staff pathologists kept up, often under difficult circumstances, with the considerable volume of routine hospital-generated work. The LCC list, then, was far from a ringing endorsement of the principle of expertise as a regular feature of inquest investigation.

46. In considering this measure the PCC was looking not only to enhance accuracy but also to lower the cost (economic and emotional) of an inquest system that, in its view, intruded unnecessarily on private grief and the public purse by investigating deaths that proved to be "natural." "Without entirely endorsing the opinion that whenever [death from natural causes] has been returned an inquest was unnecessary," the PCC's chief officer later observed, "it may fairly be held that in the great majority of such cases an efficient preliminary enquiry would have made an inquest unnecessary." LMA: LCC/PC/COR/1/1, "Coroners' Inquests—Report by the Chief Officer of the PCC," October 1901, 5. In its proposals the PCC was in a sense following in the tradition of the nineteenth-century county magistrates that had attempted to block the regular use of what they considered costly postmortem evidence by disallowing fees for those deemed "unnecessary." However, here the PCC was charting a new course for fiscal con-

servatism, harnessing postmortems as preemptive evidence to the cause of economy.

47. "The County Council and the London Coroners," *BMJ* 1 (2 Mar. 1895): 500. The *Lancet* was more direct about its possible negative implications for the ordinary practitioner. While acknowledging that the reforms were in broad accordance with the stated desires of the medical community, the *Lancet* was at pains to have the involvement of expert pathologists acknowledged as being limited to cases of "particular difficulty, and not to the general run of post-mortem examinations, which can be conducted with skill and trustworthiness by the general practitioner." Even then, the editorial continued, "the natural consequences of putting the suggested proposals in force would, in our opinion, deprive the general practitioner of a source of income [to] which he has looked forward with faith and with a sense of rightful claim." "Coroners' Inquests," *Lancet* 1 (6 Apr. 1895): 881.

48. As part of its selection process, the PCC asked each candidate the following question: "Are you prepared to entrust post-mortem examinations to a skilled pathologist, as desired by the Council, in all cases except where you are satisfied that the medical man connected with the case is competent to make a trustworthy post-mortem examination?" For a different perspective on the Troutbeck story, see D. Zuck, "Mr. Troutbeck as the Surgeon's Friend: The Coroner and the Doctors—An Edwardian Comedy," *Medical History* 39 (1995): 259–87.

49. "The Medical Man, the Coroner, and the Pathologist," *Lancet* 2 (29 Nov. 1902): 1477.

50. *Lancet* 1 (3 June 1905): 1538.

51. *BMJ* 1 (14 Mar. 1903): 644.

52. "The Post-mortem Question," *Medical Free Lance: A Weekly Journal Devoted to the Interests of the Medical Profession* 1 (1 Apr. 1905): 2. This short-lived journal (in print for less than a year) devoted a great deal of its editorial space to the Troutbeck controversy.

53. "Doctors' Bad Post-Mortems. Battersea Municipal Alliance Enlightened. Views of Local Medical Men," *Wandsworth Borough News*, 22 May 1903, 8. Emphasis original.

54. "Medical Evidence at Inquests: Deputation to the Lord Chancellor," *BMJ* 1 (16 May 1903): 1178. The other organizations represented in this joint deputation were the British Medical Association, the Medical Defense Union, and the London and Counties Medical Protection Society.

55. *Lancet* 1 (14 Mar. 1903): 758.

56. "Doctor v. Specialist: Scene at an Inquest," *Wandsworth Borough News,* 13 February 1903, 3. Emphasis original.

57. Very few of the leading manuals on postmortem practice of the day specified agency in the conduct of autopsies, tending to adopt a third-person passive construction in describing procedures for making the examination. When mortuary assistants were mentioned, the distribution of responsibility between them and the medical attendant tended to lend credence to Freyberger's claims: "If the mortuary attendant acts as an assistant," Stanley Atkinson stated, "he must not use a knife before the medical witness arrives, and each step taken must be watched by the latter." Atkinson, *The Law in General Practice: Some Chapters in Every Day Forensic Medicine* (Oxford, 1908), 47. Mortuary regulations drawn up by local authorities also suggest a flexible cooperation between pathologist and attendant. The City of London's Commission of Sewers specified that mortuary attendants "shall give every facility to members of the medical profession conducting post-mortems and render such assistance therein as practical." Corporation of London Record Office (CLRO): CSPR 27, "Duties of the Keeper of the Mortuary Buildings," 1897, rule 18. Medical men certainly paid for this assistance; noted London coroner Bentley Purchase reported in 1933 that it was usual for attendants to receive a half a crown out of every guinea of the pathologist's fee for "carry[ing] out various duties in connection with an autopsy, such as opening and sewing up the body, sawing open the skull, etc." LMA: LCC/PC/COR/2/11. On occasion, however, local authorities sought to intervene in this customary trade, causing considerable consternation on the part of the medical attendants in the process. See, e.g., the correspondence in LMA: LCC/PC/COR/3/71. For observations on the status of mortuary assistants, see Michael Clark, "Ministering to the Departed: Post-mortem Technicians and the Development of Mortuary Work as a Skilled Occupation, 1900–1980," *Bulletin of the Society for the Social History of Medicine* 40 (1987): 58–61.

58. "Notes and Comments: The Medical Man, the Coroner, and the Pathologist," *Lancet* 1 (21 Feb. 1903): 561.

59. Troutbeck to Walter Schroeder, 15 June 1905, in *Annual Report of the Coroners' Society,* 1905–6, 206–7.

60. John Troutbeck, "Modes of Ascertaining Fact and Cause of Death," *Transactions of the Medico-Legal Society* 3 (1905–6): 86–117, 98. Paul Brouardel was a leading French medicolegal expert and director of the Paris morgue, an institution that for many epitomized the centralized, intrusive, and sensationalizing nature of Continental death management. The process whereby expertise was secured only as a practical result of designating it as

such was grudgingly acknowledged even by Troutbeck's enemies, as illustrated by the following exchange between a member of the 1909 Select Committee on Coroners and W. Piercy Fox, a general practitioner slighted by Troutbeck. Fox caustically described Freyberger as a "so-called expert—not an expert recognised by the profession." When the committee member observed that Freyberger nonetheless had "very large experience?" Fox could only retort, "Well, since Mr. Troutbeck took him in hand." *British Parliamentary Papers,* 1910, "Second Report of the Coroners' Committee: Pt. 2, Evidence and Appendices," 21:603.

61. See, e.g., Troutbeck, "Inquest Juries," *Transactions of the Medico-Legal Journal* 1 (1902–4): 49–58.

62. For a brief but useful discussion of the history of overlaying, see Todd L. Savitt, "The Social and Medical History of Crib Death," *Journal of the Florida Medical Association* 66, no. 8 (1979): 853–59.

63. Troutbeck, "Modes of Ascertaining," 97. In such cases coming before his court, Troutbeck often publicly criticized local practitioners for assuming that a child had been overlaid on the basis of mere prejudice. The *Morning Leader* of 8 August 1908 reported one such case. After the local practitioner, W. Piercy Fox, testified that the dead infant had been overlaid, Troutbeck called for Freyberger's postmortem, which reported "that death was due to failure of the heart while the child was suffering from bronchitis and polypus catarrh. The usual signs of death from suffocation were all absent." In his summation, Troutbeck criticized Fox for his rash conclusions and then observed that "so many persons had been said to have suffocated their children by overlaying during the past few years that he had paid very great attention to this class of cases, with the result that it had been shown, in that district at all events, that the practice of overlaying did not exist" (5).

64. Harrington Sainsbury, "A Plea for a More Living Pathology," *BMJ* 2 (2 Oct. 1909): 928–30, 929.

65. "Mental Attitudes in Pathology," *BMJ* 2 (2 Oct. 1909): 995.

66. F. J. Smith, "Post-mortem Examinations Which Do Not Reveal the Cause of Death," *Transactions of the Medico-Legal Society* 3 (1905–6): 39–40. Smith's main objective was to demonstrate the indeterminacy of death causation and to explain why, in the face of such indeterminacy, so few examinations of cause of death ended inconclusively. To this end, Smith pointed out flaws in causal explanation derived in the clinic as well as in the postmortem room.

67. Ibid., 44.

68. "The Medical Man, the Coroner, and the Pathologist," *Lancet* 1 (28 Mar. 1903): 901.

69. For more on the importance of toxicology as a branch of nineteenth-century medical jurisprudence, see Noel G. Coley, "Alfred Swaine Taylor, MD, FRS (1806–1880): Forensic Toxicologist," *Medical History* 35 (1991): 409–27.

70. *Lancet* 1 (14 Feb. 1903): 487.

71. *Lancet* 2 (15 Sept. 1900): 817.

72. A. W. Mayo Robson, FRCS, criticized the increased reliance on laboratories that had "no personal interest in the patients from whom the samples they investigate are derived" as bad medicine and bad science. In a direct challenge to the epistemological framework within which expert pathologists assumed their superiority over the clinician at the postmortem table and the witness box, Mayo observed that "it is a generally accepted rule that a diagnosis should hardly ever be founded on any one symptom or any one method of examination . . . Yet we often see a specimen sent to a clinical laboratory without any information as to its source or of the clinical condition of the patient, and find that an opinion is expressed or a diagnosis made on the results of this examination alone" ("An Address on the Position of Pathology with Regard to Clinical Diagnosis," *BMJ* 1 [17 Mar. 1906]: 602–3).

The consequences of the shift from bedside to laboratory medicine are, of course, a central concern in the historiography of modern medicine. The classic article is N. D. Jewson, "The Disappearance of the Sick Man from the Medical Cosmology," *Sociology* 10 (1976): 225–44. John Pickstone's current work promises an important reconsideration of this shift (in his terms between "savant" and "analytical" medicine) within the broader epistemological and institutional contexts of modern science, technology, and medicine. See, e.g., John Pickstone, "Museological Science? The Place of the Analytical/Comparative in Nineteenth-Century Science, Technology, and Medicine," *History of Science* 32 (1994): 111–37.

73. F. Bushnell, letter, *Lancet* 2 (15 Sept. 1900): 840. Such complaints continued right up to the end of the period under discussion. In the words of London University's professor of bacteriology, pathologists "have too long suffered from the clinician's conception of them as purely hewers of wood and drawers of water for their particular benefit." *BMJ* 2 (26 Sept. 1925): 554. The "Cinderella" epithet comes from the discussion of this paper (559). Historians of English pathology have tended to uphold the pathologists' complaints, noting that it was not until after World War I that a viable career structure slowly began to emerge, ultimately enabling pathol-

ogy to shed its traditional image as a "way-station" to more valued work. See Maulitz, "The Pathological Tradition," 187, and Foster, "The Early History," 178–81. For a superb analysis of the tensions between technological and clinical visions of medical practice in this period, see Christopher Lawrence, "Incommunicable Knowledge: Science, Technology, and the Clinical Art in Britain, 1850–1914," *Journal of Contemporary History* 20 (1985): 503–20.

74. *BMJ* 1 (7 Feb. 1903): 344–45. From Troutbeck's perspective, of course, Badcock's protest only reinforced his contention that experts were needed to avoid precisely this kind of private knowledge that reflected (as in the example of overlaying) prejudice rather than knowledge.

75. *BPP*, 1909, 15:401. An earlier correspondent to the *BMJ* had shared Little's concern: Preemptive specialist postmortems were in his view "unreasonable and unnecessary on either legal or scientific grounds, and without being in any way sentimental the public would justly express abhorrence at such mutilation of the dead." The coroner's court, he added in conclusion, "should not be a pathologist's lecture room." *BMJ* 2 (16 July 1904): 152. Proponents of an expert-centered inquest saw the innovation of preliminary postmortems as protecting against a different form of intrusive curiosity, that of an excessive publicity entailed at inquests that exposed private grief to the public gaze. See, e.g., William A. Brend, "The Necessity for Amendment of the Law Relating to Coroners and Inquests," *Transactions of the Medico-Legal Society* 10 (1912–13): 143–97, 180–83.

76. W. Wynne Westcott, "Twelve Years' Experience as a London Coroner," *Transactions of the Medico-Legal Society* 4 (1906–7): 15–32, 20.

77. Discussion following Brend's "The Necessity for Amendment," 193.

Chapter 5: Fatal Exposures

1. Corporation of London Record Office (CLRO), Southwark inquests, 1884, no. 38. Simpson's was one of several cases in the 1880–90s heard before the coroner for the City of London and Southwark in which dread of chloroform was mentioned. See also Elizabeth Sole (Southwark inquests, Apr. 1882, no. 75); Frederick Whitehouse (Southwark inquests, June 1886, no. 145); Frances Louisa Slade (Southwark inquests, June 1890, no. 121); and Rebecca Hollington (City of London inquests, Sept. 1896, no. 140).

The influence of the patient's subjective state on the outcome of a surgical procedure was a part of the discourse on anesthetics almost from the outset. In the early decades of its use, the tendency was to regard anesthesia as a means of liberating the surgical patient from potentially mortal fear. Citing ancient and contemporary authority, T. B. Curling, in his 1848 address to the Hunterian Society, maintained that "fear is the strongest, as well as the most painful, of all the passions, and when excited in a high degree . . . it powerfully depresses the vital powers, enfeebles the whole frame, and even produces death." The recent discovery of anesthesia, he continued, had provided an antidote to this "deadly poison." Curling, *The Advantages of Ether and Chloroform in Operative Surgery* (London, 1848), 16, 18. Yet the opposite configuration—anesthesia as a stimulant to mortal fear—was also articulated in the early history of anesthetics. "I cannot help, in fact, believing that emotional depression so curiously and inextricably interwoven with the functions of the spinal and sympathetic system had much to do with these mysterious deaths," Charles Kidd wrote of anesthetic fatalities in 1858, observing that in Paris, consequently, "a great deal of trouble is taken by many *petits soins* to calm the emotions and assure patients of perfect safety before they begin to take chloroform." Charles Kidd, *On Ether and Chloroform as Anaesthetics*, 2d ed. (London, 1858), 30.

2. This was not the first time that hospitals had crossed paths with the inquest, of course. One of Wakley's main claims for a medical coronership, after all, had been the need for competent and independent monitoring of the "Bats and Corruptionists"—his regular epithet for the London hospital establishment of his day. This world, in his view, was the medical analogue of "Old Corruption," and he used the editorial pages of the *Lancet* and later (though to a lesser extent) the Middlesex coroner's court as platforms for exposing its failures. The use of inquests in cases of anesthetic death in the last quarter of the century was quite different. Here the issue was not the legitimacy of the medical system itself, but a worrisome and persistent flaw within a system whose legitimacy was for the most part taken for granted. It was the very targeted nature of anesthetic inquests, their pretension to pass judgment on an integral part of ordinary medical practice, that made them such an irritant to those who believed that the battles for professional self-determination had (or should have) already been won.

3. My consideration of anesthesia is limited to inhalation anesthesia in the context of English hospital practice and excludes the use of local anesthetics, the use of anesthetic agents in nonhospital settings like dental surgeries and bone-setting clinics, and the development, toward the end of the period under consideration, of alternative methods of delivering general anesthesia (e.g., spinal injection).

4. Cited in A. J. Youngson, *The Scientific Revolution in Victorian Medicine* (New York, 1979), 80. Youngson considers the "revolutionary" effect of anesthesia in ch. 3. For an important discussion of the early English

experience with chloroform and ether, see Barbara Duncum's comprehensive study, *The Development of Inhalation Anaesthesia, with Special Reference to the Years 1846–1900* (London, 1947). A rich analysis of the gendered doctor-patient relationship as mediated by the introduction of chloroform can be found in Mary Poovey, *Uneven Developments: The Ideological Work of Gender in Mid-Victorian England* (Chicago, 1988), ch. 2. Martin S. Pernick's *A Calculus of Suffering: Pain, Professionalism, and Anesthesia in Nineteenth-Century America* (New York, 1985) is a stimulating discussion of the place of anesthesia in American medical practice and culture.

5. Chloroform deaths by far outstripped those involving ether, and it was soon considered the more dangerous of the two agents. It was also more widely used than ether. Over the course of the nineteenth century, there was intense debate between proponents of chloroform and ether. Chloroform was universally known to be easier to administer, less unpleasant for the patient, and capable of producing a deeper anesthetic state, thus improving surgical conditions. Champions of chloroform maintained that ether victims died more frequently from surgical shock due to their insufficiently anesthetized condition, a fact that they held negated ether's claims as the more benign of the two agents.

6. There was, for example, no statutory requirement to notify the coroner of deaths under anesthesia, to mention on a death certificate anesthetics as a contributing cause, or for a local registrar of death to forward any such certificate to the coroner's office. The General Register Office (GRO) thus depended primarily on inquest verdicts that made mention of an anesthetic agent for its yearly figures on anesthetic fatalities. Arrangements for a formal notification procedure were not made until after the passage of the 1926 Births and Deaths Registration Act (16 & 17 Geo. 5, c. 48). Crisp's figures were published in the *Lancet* 1 (4 June 1853): 523; Snow's are cited in Youngson, *The Scientific Revolution*, 80.

7. *Lancet* 1 (8 Feb. 1890): 317.

8. The numbers of deaths reported to the GRO from 1870 to 1898 were 16 in 1870, 26 in 1880, 41 in 1890, 78 in 1895, 126 in 1897, and 90 in 1898. A *British Medical Journal* correspondent contrasted these with earlier figures, observing that between 1848 and 1863 a total of 86 deaths had been reported. *BMJ* 1 (12 Mar. 1898): 704–5. Again, there were no generally accepted figures on death rates. In his 1890 letter complaining about this situation, Roger Williams offered the results of his own study of the Barts's register in an effort to address this gap, reporting that in 1878–87 the hospital performed 12,368 chloroform administrations with 10 deaths

(1 in 1,236) and 14,581 ether administrations with 3 deaths (1 in 4,860). In 1902 Williams updated his findings: in 1891–1900, Barts administered chloroform 23,452 times with 20 deaths (1 in 1,172) and ether 15,935 times with no deaths. *Lancet* 1 (8 Feb. 1890): 317; 1 (7 June 1902): 1643. According to Anne Dally, the most common estimates for anesthetic death in this period ranged from 1 in 2,000 to 1 in 3,000 cases. Anne Dally, "Status Lymphaticus: Sudden Death in Children from 'Visitation of God' to Cot Death," *Medical History* 41, no. 1 (Jan. 1997): 70–85.

9. *Inexplicable* was the term used by the 1909 Departmental Committee on Coroners in its special report on anesthetic fatalities. *British Parliamentary Papers* (hereafter *BPP*), 1910, vol. 21, "Report of Inquiry into the Question of Deaths Resulting from the Administration of Anaesthetics," 787.

10. *BMJ* 1 (12 Mar. 1898): 704.

11. Dudley W. Buxton, *Anaesthetics: Their Uses and Administration* (London, 1888); J. Frederick Silk, "Anaesthetics a Necessary Part of the Curriculum" (paper read before the Thames Valley Branch of the British Medical Association [BMA]), *Lancet* 1 (28 May 1892): 1178–80. Silk, an anesthetist attached to Guy's and Royal Free Hospitals, was the founder of the Society of Anaesthetists.

12. The Society of Anaesthetists was formed in London on the initiative of practicing hospital anesthetists and had an initial membership of forty. It began publishing its *Transactions* in 1898 and in 1908 became associated as a section of the Royal Society of Medicine.

13. Silk, "Anaesthetics a Necessary Part," 1178. Nor did the situation change quickly. The General Medical Council (GMC) rejected a representation of the Society of Anaesthetists in 1901, asking that anesthetic instruction "be compulsorily included as a separate subject of the medical curriculum." In 1907 the GMC reversed itself and made a formal request that all examining bodies require evidence of instruction, but a survey undertaken by the British Association for the Advancement of Science later that year found that only eight of twenty-two medical examining bodies in the United Kingdom required evidence of instruction (Digby Coates-Preedy, in F. W. Hewitt, *Anaesthetics and Their Administration: A Text-Book for Medical and Dental Practitioners and Students,* 4th ed. [London, 1912], 652). According to Duncum, it was not until 1911 that the GMC could report that every teaching body in Great Britain was following its recommendations (Duncum, *The Development,* 531). Even then, having recognized anesthetists on staff did not mean that in practice anesthetics had been passed on to specialists. At Guy's Hospital, for example, only 30% of the

administrations recorded in 1910 were done by staff anesthetists; a decade later the rate was 35%. London Metropolitan Archives (LMA): H9/GY/A264/1–2, Guy's Hospital Yearly Returns of Anaesthetics, 1908–12, 1913–20.

14. *Annual Report of the Coroners' Society,* 1894–95, 29–30.

15. Ibid., 1902–3, cited in *BMJ* 2 (2 July 1904): 48–49.

16. *BPP,* 1910, vol. 21, "Second Report of the Coroners' Committee: Pt. 3, Evidence and Appendices," 656. Even the most implacable foes of public inquiry into anesthetic death had to agree with the substantive point that anesthetics were poisons. Indeed, from 1891 the registrar general's annual reports classed these deaths according to the following taxonomic order: "Violent Deaths; Deaths from Accident and Negligence; Poisons and Poisonous Vapours." Disagreement lay not in the association of anesthesia with poison, but rather in the significance to be attributed to this basic fact.

17. *BPP,* 1909, vol. 15, "First Report of the Departmental Committee Appointed to Inquire into the Law Relating to Coroners and Coroners' Inquests: Pt. 2, Evidence and Appendices," 519.

18. *BPP,* 1910, 21:673. Troutbeck played a significant role in the controversy surrounding hospital inquests, with his high-profile battles with the controversial surgeon Victor Horsley being cited as one of the reasons for commissioning a parliamentary inquiry in the first place. Unfortunately, I cannot go into the details of the Troutbeck-Horsley confrontation here; suffice to say that it followed the contours of the broad disputes discussed in this chapter, albeit in a characteristically intensified fashion. For the basics of the story, see D. Zuck, "Mr. Troutbeck as the Surgeon's Friend: The Coroner and the Doctors—An Edwardian Comedy," *Medical History* 39 (1995): 259–87.

19. "The Mania for Operations," *Lancet* 1 (11 Jan. 1851): 54.

20. F. W. Hewitt, "The Past, Present, and Future of Anesthesia," *Practitioner* 57 (1896): 348.

21. *Lancet* 1 (11 Apr. 1908): 1107.

22. *Lancet* 1 (11 Jan. 1851): 54.

23. Among the earliest charges levied against anesthetists (chloroformists in particular) was that of taking carnal advantage of their unconscious female patients. The charge was quickly incorporated into the canon of medicolegal writing on anesthesia; on the one hand, the anesthetic agent was imputed with hallucinogenic powers exciting the sexual imagination of female subjects; on the other hand, anesthetists were encouraged to administer only in the presence of a third party. See Poovey, *Uneven Developments,* ch. 2, and Youngson, *The Scientific Revolution,* ch. 3.

24. Hewitt, *Anaesthetics and Their Administration*, 643–44. Handwritten notes made by Southwark deputy coroner A. Langham for an 1893 case of anesthetic death at Guy's Hospital revealed a practical concern with the questions raised in Coates-Preedy's theoretical discussion: "*Anaesthetics* given to allay pain and prevent exhaustion; no right to perform operation ag. the Will of the Patient *so long as patient preserved consciousness and will.*" CLRO: Southwark inquests, no. 182, 4 September 1893. Emphasis original.

25. Occasionally the "excitement" stage was described from within. L. G. Guthrie's Oxford University MD thesis contains the following account of an anesthetist's own experience with the agent:

> After a few minutes I felt suddenly that I was being suffocated and tried to make signs that this was so, but all to no purpose. Then, whether struggling or not I cannot tell, I felt that I did not move hand or foot and gave myself up for lost. At this moment I felt the apparatus removed from my face, and I shall never forget the sensation of that delicious breath of fresh air. It was like coming to the surface after a long swim under water. I remember nothing more till I woke to find the operation over. But I resolved that even if, as in my own case, such a method of administration might be safe, still I would endeavour to save my patients from such agony of impending death as I had felt.

L. G. Guthrie, "Chloroform Narcosis in Children," 1893, 61, library of the Royal College of Physicians, London.

26. The best account of the movement in England is still R. D. French's *Antivivisection and Medicine in Victorian Society* (Princeton, 1975). Antivivisection societies like Francis Power Cobbe's Victoria Street Society produced a prodigious quantity of literature in their attack on all aspects of modern medical experimentalism. According to French, in one year alone Cobbe's organization put into circulation over eighty-one thousand books, pamphlets, and leaflets. French, *Antivivisection and Medicine*, 255.

Anesthesia itself played a complex role in the debates over animal experimentation, driving a wedge between antivivisectionists over the question of whether painless experiments were humane and dividing the experimental community on the usefulness of results derived from such "denatured" subjects. For more on this issue, see Stewart Richards, "Anaesthetics, Ethics, and Aesthetics: Vivisection in the Late Nineteenth-Century British Laboratory," in Andrew Cunningham and Perry Williams, *The Laboratory Revolution in Medicine* (Cambridge, 1992), 142–69. Other notable discussions of antivivisectionism include the essays in Nicholaas A. Rupke, ed., *Vivi-*

section in Historical Perspective (London, 1987), esp. Mary Ann Elston's
"Women and Anti-vivisection in Victorian England, 1870–1900," 259–94,
and Coral Lansbury, *The Old Brown Dog: Women, Workers, and Vivisection in Edwardian England* (Madison, Wis., 1985). For the movement's relation to the broader context of Victorian humanitarian crusades, see Brian
Harrison, *Peaceable Kingdom: Stability and Change in Modern Britain*
(Oxford, 1982), ch. 2.

27. "Human Vivisection," *Daily Chronicle*, 15 May 1894. The *Chronicle*'s editorial was one in a series devoted to the scandal at the Chelsea Hospital for Women, where high operative mortality rates were widely attributed to the experimental enthusiasm of a staff led by the notorious obstetric
surgeon Spencer Wells.

28. E. A. King, "Deaths under Chloroform," *Nineteenth Century* 43
(Mar. 1898): 515–20. The terms "smothering" and the "School of Stiflers"
come from her "Death and Torture under Chloroform," *Nineteenth Century* 43 (June 1898): 985–93, 989, 990.

29. King, "Death and Torture," 992.

30. "The Truth about Hospitals," *Verulam Review* 6 (1896–97): 358.

31. Ibid., 362.

32. "Half-truths are notoriously dangerous," Buxton explained, "and
are peculiarly liable to occur in the pages of the lay press which deal with
the recondite questions of science." Buxton, "Death under Chloroform: A
Reply," *Nineteenth Century* 43 (Apr. 1898): 668–75, 668. King rejected the
view that internal discussion was the proper way of handling matters of this
nature. The structural problem, she insisted, lay not with the generic requirements of the lay press, but rather with the constrained nature of
intraprofessional discussion. Buxton's "reply," devoted almost wholly to
the esoteric dispute concerning the physiology of anesthetic death, fit seam-
lessly into an existing web of inconsequential professional discourse: "A
perpetual domestic war on the subject is decorously waged in the medical
press without either side obtaining a decisive victory." Professional "airing"
was merely a form of containment, one that missed the essential matter of
public interest. King, "Death and Torture," 985.

33. "Death Certification: Pt. 3," *BMJ* 2 (8 Dec. 1900): 1649.

34. Frederick Treves, "Anaesthetics in Operative Surgery," *Practitioner*
57 (1896): 377–83, 378–79. Treves, a staff surgeon at the London Hospital, is better known to modern readers as the medical attendant and ex-
hibitor of John Merrick, "the elephant man."

35. Hewitt, *Anaesthetics and Their Administration*, 18–19, 263–64.

36. "Coroners' Inquests on Deaths under Anaesthetics," *Lancet* 1 (31 Jan. 1914): 327.

37. This was, of course, part of a wider contemporary critique of "sensationalism" in modern mass culture, in which "the new journalism"—described by one historian as a shift both in content (typified by "the transformation of the colourless report to the human interest story") and form ("a typographical revolution in which the paragraph superseded the column and the headline overshadowed the paragraph, and with 'cross-heads' to break up solid masses of print and facilitate rapid reading")—occupied a key place. Alfred F. Havinghurst, *Radical Journalist: H. W. Massingham (1860–1924)* (London, 1974), 18. Historians and critics have traditionally viewed this as a cooptive transformation of a previously more independent and politicized popular press into what Raymond Williams has described as "a highly capitalised market product for a separated 'mass' readership." Williams, "The Press and Popular Culture: An Historical Perspective," in George Boyce, James Curran, and Pauline Wingate, eds., *Newspaper History from the Seventeenth Century to the Present Day* (London, 1978), 49. See also Virginia Berridge, "Popular Sunday Papers and Mid-Victorian Society," in Boyce, Curran, and Wingate, *Newspaper History;* Alan J. Lee, *The Origins of the Popular Press in England, 1855–1914* (London, 1976); and Joel Wiener, *Papers for the Millions: The New Journalism in Britain, 1850s to 1914* (New York, 1988). Recent attention to the symbolic framework within which the late Victorian popular press operated has led to a more dynamic interpretation of this phenomenon. See esp. Patrick Joyce, *Democratic Subjects: The Self and the Social in Nineteenth-Century England* (Cambridge, 1994), chs. 14 and 16; Judith Walkowitz, *City of Dreadful Delight: Narratives of Sexual Danger in Late Victorian London* (Chicago, 1992), chs. 3 and 4; and Rowan McWilliam, "The Mysteries of G. W. M. Reynolds: Radicalism and Melodrama in Victorian Britain," in Malcom Chase and Ian Dyck, eds., *Living and Learning: Essays in Honour of J. F. C. Harrison* (Aldershot, 1996), 182–98.

38. *Lloyd's Weekly Newspaper,* 8 December 1907, 5.

39. *The Times,* 13 December 1910, 4.

40. *The Times,* 15 December 1910, 4.

41. "The Responsibility of Our Hospitals and Deaths under Anaesthetics," *Lancet* 2 (1 Oct. 1910): 1023.

42. "Coroners' Inquests on Deaths under Anaesthesia," *Lancet* 1 (31 Jan. 1914): 328.

43. Royal Society of Medicine Archives: Minute Book (K1), Section on Anaesthetics, Council Minutes, "Draft Memorial of the Section to the

Coroners' Society," 10 March 1913. This document was drafted by the section in an attempt to convince coroners of the need to curtail their inquests.

44. The phrase "anaesthetic terror" is from Buxton, "Sleep and Her Twin Sister, Death," in *Contemporary Review* 103 (Jan. 1913): 103–8, 108.

45. "Chloroform and Its Dangers," *Lancet* 1 (15 Jan. 1870): 90. The *Lancet* stated its suspicions more forthrightly at the end of the century, when it ventured that "probably the unsatisfactory paragraphs so often seen in the daily press add not a few victims to the list of casualties under anaesthetics." *Lancet* 1 (17 Jan. 1896): 1093.

46. Silk's discussion of Ludwig Freyberger, "Seventy-four Cases of Sudden Death under Anaesthetics," *Transactions of the Medico-Legal Society* 5 (1907–8): 20–77, 72.

47. Ibid., 66.

48. *BPP*, 1910, 21:700. Horsley was a leading London neurosurgeon, the longtime head of the BMA's Medico-Political Section, and a prominent member of the Medical Defense Union. His unrepentant advocacy of the rights of medical experimentation won him the public enmity of the antivivisectionist campaign.

49. Royal Society of Medicine Archives: "Draft Memorial," 1913. Even those sympathetic to the inquest saw overpublicity as a problem. The Birmingham coroner, E. A. Gibson, stated in testimony before the 1909 parliamentary commission that he regularly appealed to newspaper reporters to exercise restraint: "I say: 'I am persuaded, as a medical man myself, and having had considerable experience, that it would be to the disadvantage of patients to have their nerves disturbed, and they are sure to remember those newspaper reports when their turn comes to have to go under an anesthetic.'" "If a man thinks he is going to die it predisposes him to die?" a member asked, to which Gibson replied, "Undoubtedly." *BPP*, 1910, 21:656.

50. The anesthetist H. Bellamy Gardner observed that the public read inquest reports with "an unreasoning fear," which led it to ignore the fact that often the anesthetic was found not to be the cause of death. Analogizing beyond the instance of surgical anesthesia, Gardner cast his opposition as a matter of common humanity: "If a patient is dying from heart disease in a hospital ward humanity leads us to screen him off from view, not to call aloud to other patients, 'Look what may happen to you as a result of rheumatic fever.'" *Lancet* 1 (11 Apr. 1908): 1107.

51. *BPP*, 1909, 15:450.

52. Ibid.

53. Ibid., 449.

54. For a discussion of Fletcher and the work of the Medical Research Council, see Joan Austoker, "Walter Morely Fletcher and the Origins of a Basic Biomedical Research Policy," in Joan Austoker and Linda Bryder, *Historical Perspectives in the Role of the MRC* (Oxford, 1989), 23–33.

55. Sir Walter Fletcher to Sir Arthur Robinson (Ministry of Health [MH]), 17 August 1921, Public Record Office (PRO): MH58/264.

56. PRO: MH58/264, correspondence from Middlesex Hospital (4 Nov. 1920); University College Hospital, London (5 Nov. 1920); and St. Thomas's Hospital (10 Dec. 1920). The frustration of these hospital officials at the intrusion of the inquest was echoed in journals and professional societies. The Section on Anaesthetics of the Royal Society of Medicine argued in its committee report to the Ministry of Health on anesthetic death that the current procedure of hybrid inquiry was the most significant impediment to increased knowledge of an anesthetic death: "There is little or no possibility of this increase under the existing arrangements for post-mortem examination." As an opinion piece in the *British Journal of Anaesthesia* later observed, "There is a gross waste of valuable clinical and pathological material which, properly co-ordinated, might be of great scientific value." Royal Society of Medicine Archives: *Minute Book* (K1), Section on Anaesthetics, "Report of the Committee on Deaths from Anaesthesia," January 1927, 254; "Coroners and Anaesthetists," *British Journal of Anaesthesia* 9, no. 2 (Jan. 1934): 80.

57. Home Office to Ministry of Health, 18 January 1922, PRO: HO45/15558/193114/54.

58. "Deaths under Anaesthetics," *Lancet* 1 (5 May 1923): 910.

59. *BPP*, 1910, 21:700.

60. *BPP*, 1909, 15:520–21.

61. In *Annual Report of the Coroners' Society,* 1922–23, 208.

62. "The Use of Anaesthetics," *Lancet* 2 (15 Sept. 1888): 524. Jennifer Beinhart's survey of twentieth-century newspaper accounts of anesthetic inquests corroborates the *Lancet*'s observation. Beinhart characterizes these as formulaic, "their inevitable concluding sentences reporting verdicts of accidental death, and recording the coroner's opinion that the anesthetic had been properly administered." Jennifer Beinhart, *A History of the Nuffield Department of Anaesthetics, Oxford, 1937–87* (Oxford, 1988), 66.

63. "Operations under Anaesthetics," *Law Journal* 43 (11 Jan. 1908): 16. The response of the BMA's deputy medical secretary to an anesthetist's question in 1936 indicates that to that date there had been no case known to the association of an anesthetist being brought up on manslaughter

charges (Wellcome Institute Contemporary Medical Archive Collection: SA/BMA/E/3).

64. Henry Robinson, *Hewitt's Anaesthetics and Their Administration* (London, 1922): 553.

65. Some contemporaries were not reluctant to put forward this explanation. Elizabeth King complained that in anesthetic inquests "it is the patient's heart that was always in fault. The coroner's jury will be guided by the evidence given by himself and the other professional men or nurses present, whose statements are fettered by professional etiquette, and it will exonerate him from all blame in the matter." She was convinced, on the other hand, that if juries "occasionally returned a verdict of manslaughter, the reign of the stiflers would not last long." King, "Death and Torture," 990–91.

66. The most interesting of the organic-idiosyncratic explanations was the condition described as status lymphaticus, which represented the apogee of the pathologist's claim to be able to locate a tangible cause of anesthetic death. Status lymphaticus placed the diffuse notion of idiosyncrasy within the purview of the pathologist by tying the idea of a constitutional predisposition to anesthetic death to a pathologically recognizable cause, an enlarged and persistent thymus. The doctrine of status lymphaticus was the subject of considerable controversy over its thirty-year life span in the context of English anesthetics (roughly 1907–31). Critics were numerous and outspoken in their dismissal of the condition's very existence. Walter Fletcher, for one, informed the Ministry of Health that "there is good reason to think that this condition does not really exist at all, and is only a pompous name brought forward to hide ignorance." Walter Fletcher to Sir Arthur Robinson, 12 May 1920, PRO: MH58/264. Yet, starting in 1913, status lymphaticus was given its own subheading in the registrar general's tables on anesthetic death and was regularly invoked to account for cases of sudden death under anesthesia, especially of young children, by prominent pathologists like Spilsbury and the police surgeon Dr. Rose. The Medical Research Council and the Pathological Society responded to the controversy by forming a joint commission in the early 1920s to determine the legitimacy of the condition. Its findings, issued in 1931, pronounced that "there is no evidence that so-called 'status-thymico lymphaticus' has any existence as a pathological entity." For a discussion of the curious history of status lymphaticus, see Dally, "From 'Visitation of God' to Cot Death."

67. Such consensus as there was existed on the level of professional pragmatism and not on theoretical grounds. For an incisive account of the state of research into anesthetic death at the beginning of the twentieth

century, see Christopher Lawrence, "Experiment and Experience in Anaesthesia: Alfred Goodman Levy and Chloroform Death, 1910–1960," in Lawrence, ed., *Medical Theory, Surgical Practice: Studies in the History of Surgery* (London, 1992), 263–93.

68. "The Coroner's Catechism" was for guidance only and had no prescriptive force. It, however, was reprinted in subsequent editions of the standard work on medical jurisprudence, was circulated by the Coroners' Society to its members, and was generally held to have been followed by at least the London coroners.

69. The London Hospital's Leonard Hill, a leading figure in London experimental physiology, alluded to one of the ancillary consequences of this situation for the practicing anesthetist: "Chloroform is a drug used by the young anaesthetist with the utmost hardihood, and until he has had the misfortune in his practice to meet with a death caused by it, he derides the danger of the drug, and asserts that its safety merely depends on the care and skill of the administrator. After losing his patient, he falls to descanting on the unavoidable dangers of the drug, dangers which he is now the first to maintain cannot be met by any degree of skill in administration." Cited in John Stuart Ross, *Handbook of Anaesthetics* (Edinburgh, 1919), 120.

70. John Snow, "On the Fatal Cases of Inhalation of Chloroform," in Albert Faulconer and Thomas E. Keys, eds., *The Foundations of Anaesthesiology* (Springfield, Mass., 1965), 481.

71. *BPP*, 1909, 15:449.

72. Ibid., 15:596.

73. *BPP*, 1910, 21:590.

74. Buxton, *Anaesthesia: A Clinical Study* (Dublin, 1892), 6.

75. A. G. Vernon Harcourt to A. D. Waller, 12 December 1908, PRO: HO45/10550/162583/38. Harcourt was an Oxford physical chemist and a leading proponent of the "dosimetric" approach to chloroform induction. Waller was a lecturer in physiology at St. Mary's and head of the BMA's chloroform commission.

76. As a later critic of the value of inquests into anesthetic deaths bluntly put it, "Advantage is taken of the common lay belief that a post-mortem necessarily discloses the cause of death." R. R. MacIntosh, "Deaths under Anaesthesia," *British Journal of Anaesthesia* 31 (1948–49): 129.

Epilogue

1. Coroners (Amendment) Act, 1926 (16 & 17 Geo. 5, c. 59), s. 20 (1 & 2).

2. *BPP*, "Judicial Statistics, England and Wales: Pt. 1, Criminal Statis-

tics: Coroners' Returns to the Home Office," 1928, 25:138; 1930–40, 11:472.

3. Public Record Office (PRO): HO45/23983/138729.

4. Inquiries undertaken by the Home Office in the early 1930s revealed persisting ad hoc arrangements in local sanitary districts, described by one parliamentarian as "scandalous." PRO: HO45/17009/533786/39. In 1946 the *Lancet* was still looking forward to a day when "pathologists would no longer have to work, as they are sometimes expected to do, in sheds adjoining pig-swill plants or knacker's shops, in frigid barns, or in blacked-out corporation yards." "Autopsies for the Coroner," *Lancet* 1 (16 Mar. 1946): 388.

5. Medical examination of the body could stand in for a full public inquiry only if the coroner was convinced of the death's natural and uncontroversial nature. Violent or otherwise unnatural deaths or deaths taking place in prisons or in any of the other spaces listed in prior acts as mandating an inquest could not be handled on the basis of medical findings alone [s. 21(3)]. A similar set of guidelines obtained for nonjury inquests, including a stipulation mandating juries when "death occurred in circumstances the continuance or possible recurrence of which is prejudicial to the health or safety of the public or any section of the public." This category, in the view of the 1927 editor of *Jervis on Coroners,* would include epidemics, outbreaks of food poisoning, and deaths under anesthetics. F. Danford Thomas, *Sir John Jervis on the Office and Duties of Coroners,* 7th ed. (London, 1927), 135. The full list of restrictions on nonjury cases was given in s. 13(4).

6. Air-raid deaths and deaths resulting from munitions-works explosions were held in closed court. Home Office file-notes indicate a further restriction to which coroners acceded: "Coroners refrain from holding inquests in cases where soldiers have died from wounds received in battle, and . . . meet the views of the Admiralty by holding inquests without a jury in some other cases" (Herbert B. Simpson memo, 7 May 1917, PRO: HO45/10815/313374/17).

7. PRO: HO45/10815/313374/14, 25 April 1917. Original emphasis.

8. PRO: HO45/12605/367567/2, 9 August 1918, Herbert B. Simpson file note. For parallel suggestions about the effect of the Great War on public interest in death in England, see David Cannadine, "War and Death, Grief and Mourning in Modern Britain," in Joachim Whaley, ed., *Mirrors of Mortality: Studies in the Social History of Death* (London, 1981), 187–219. There is also a large and growing literature on the war and the memo-

rialization of the dead. See, e.g., Jay Winter, *Sites of Memory, Sites of Mourning* (Cambridge, 1995), esp. pt. 1.

9. PRO: HO45/12605/367567/2, Herbert B. Simpson memorandum, undated (c. Aug. 1918). Concern over the possible damaging effects of secrecy was echoed even in departments less directly engaged in the politics of transparency. In a memo urging "control" of inquests arising out of accidents at munitions works, an official from the Ministry of Munitions nonetheless conceded the risk that "a policy of absolute exclusion would provoke criticism and stimulate local rumours." Michael Heseltine, 29 January 1917, PRO: HO45/23968/A47022/16.

10. PRO: HO45/10336/138769/1, 1906 Home Office memo.

11. PRO: HO45/12605/367567/146, Herbert B. Simpson memo, 7 February 1910.

12. PRO: HO45/11214/403923/8, "Law as to Coroners and Inquests" (Home Office paper, Oct. 1920), 9.

13. PRO: HO45/11214/403923/29, 1923.

14. PRO: HO45/11214/403923/8, Arthur Locke docket memo (23 May 1921), re the Home Office, Ministry of Health, and General Register Office Conference on Death Certification and Coroners.

15. PRO: HO45/13998/563027/1, extract from Waldo's 1928 Annual Report to the Corporation of London, in a memorandum by Arthur Locke, 25 June 1930.

16. PRO: HO45/12285/453044/2, 1 February 1924.

17. PRO: HO45/23968/A47022/28, 3 October 1929.

18. "Report of the Committee on Death Certification and Coroners" (Her Majesty's Stationers Office, cmnd. 4810, 1971), 133–34.

Index

Numbers in *italics* indicate illustrations.